BRIAN ALLAWAY is a retired firefighter. In 1969, at the age of 16, Brian joined the Belfast Fire Brigade and served in Northern Ireland during the Troubles. In 1994, he transferred to Lothian and Borders Fire Brigade in Edinburgh and was promoted to Firemaster in 2002, the last person to hold that title. He has a BA from the Open University, an MSSc from the Queen's University of Belfast and a PhD from The University of Edinburgh. His first book, *Culture, Identity and Change in the Fire and Rescue Service*, was published in 2011 by the Institute of Fire Engineers. A Fellow of the Institution of Fire Engineers and recipient of the Queen's Fire Service Medal, Brian lives in Edinburgh.

D1380372

Firefighters of Belfast:

The Fire Service During the Troubles

1969–1994

BRIAN ALLAWAY

Luath Press Limited

EDINBURGH

www.luath.co.uk

First Published 2018

ISBN: 978-1-912147-45-8

The author's right to be identified as author of this book under the
Copyright, Designs and Patents Act 1988 has been asserted.

The paper used in this book is recyclable. It is made from low chlorine
pulps produced in a low energy, low emission manner from renewable forests.

Printed and bound by Bell & Bain Ltd., Glasgow

Typeset in 11 point Sabon by Lapiz

Photographs, unless otherwise indicated, reproduced with the kind
permission of the Northern Ireland Fire and Rescue Service

Dedicated to the firefighters of Belfast, past, present and future.

Contents

Foreword

THE PERIOD OF HISTORY that is the subject of this book was an important time for the Northern Ireland Fire Brigade. Because of the situation in Northern Ireland, many of those who served at this time were involved in incidents that were unique for a United Kingdom Fire Service. Much of what happened during these difficult days was horrific, and this book tells the story of events seen through the eyes of the Brigade. I believe this book also provides an insight into the humanity and spirit of the firefighters, as well as the impact on them at a personal level.

When I joined the Fire Brigade in the 1980s, I learned about my new career in a number of ways. There was the formal teaching environment that I experienced during my training, but probably much more effective (and longer lasting!) was the wealth of information that came to me as a young fireman through the stories and traditions that were a part of everyday life. The list of contributors that appears in the early pages of this book reads like a nominal roll from my early career. These people were my educators and role models; their experiences were passed on as part of the folklore and the culture of the Fire Brigade.

This book contains a record of the events of that time from a Fire Brigade's perspective, and for me the real attraction is that it combines these records with first hand accounts from people that were involved, bringing these significant events to life in the words of the people who lived through them.

As I approach the end of my career, I am conscious that I am just the latest in a line of fire fighters that have had the huge privilege of leading an amazing group of people, and I reflect on the fact that the Fire & Rescue Service of today has changed in many ways. However, there is much that remains the same. The selfless dedication, the courage and the calm 'can-do' attitude is still prevalent among the young firefighters, just as it has always been.

The Northern Ireland Fire & Rescue Service of today looks to the future and we continue to be very proud of where we have come from and our unique history. It is my pleasure to commend Dr Brian Allaway on his work and to recommend it to all those who have an interest in an amazing organisation at a unique time in its history.

Dale Ashford QFSM.
Chief Fire and Rescue Officer, Northern Ireland Fire Brigade (retired).

Acknowledgements

I HAVE VALUED THE help, advice and encouragement of many people during the development of this book and I would like to place on record my thanks to all of them. In particular I must thank the contributors, who are listed prior to the introduction, for the time and memories they so freely gave to me. It was a great pleasure to meet up with them, share experiences, have a beer and often a laugh, fight old fires and 'swing the lights' a little. Memory can be a fickle thing and often the contributors did not remember specific dates or addresses. However, I have done my best to corroborate everything that is in this book and any mistakes are, of course, mine alone. I am grateful to Mags McKay, who did the brilliant job of putting together the map of Belfast and making the chronology look better than I could, and to my publishers, Luath Press, particularly Alice Latchford, who made many helpful suggestions to improve the initial manuscript. I must also thank the Belfast Corporation, Fire Authority for Northern Ireland and Lothian and Borders Fire and Rescue Board for employing and developing me during my 41 years in the fire and rescue service. They also provided systems which allowed their employees to enhance their education, and I am just one of many who are grateful for the opportunities this gave to those of us who started our careers with little in the way of formal qualifications. I am also indebted to the many members of the fire and rescue services with whom I have served and enjoyed a great comradeship, and not a few laughs, despite some of the difficult situations we had to deal with. I am sure that it is due in no small measure to these people that I am who I am. Last but not least I must thank my children, Kerry and Steven, and grandchildren, Luke, Amy, Hayley, Abbie, Dylan and Matthew, and in particular my wife Diane, all of whom I care for deeply and who all seem to put up with me. Diane also deserves the credit for her magnificent proof reading and for correcting my terrible grammar and spelling! (It's a little known fact that when I left Red Watch Ardoyne, among the parting gifts they gave me was a rubber and a dictionary).

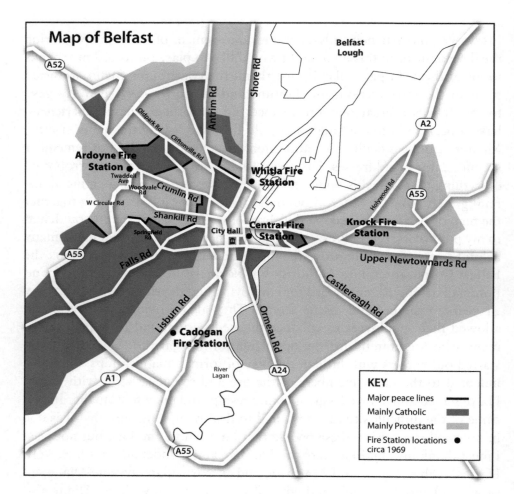

Map of Belfast

A52

Belfast
Lough

A2

Oldpark Rd

Antrim Rd

Shore Rd

Cliftonville Rd

Ardoyne Fire
Station

Twaddell
Ave

Whitla Fire
Station

Woodvale
Rd

Crumlin Rd

Holywood Rd

A55

W Circular Rd

Shankill Rd

Knock Fire
Station

Springfield
Rd

City Hall

Central Fire
Station

A55

Falls Rd

Upper Newtownards Rd

Lisburn Rd

Castlereagh Rd

Cadogan
Fire Station

Ormeau Rd

A1

River
Lagan

A24

A55

KEY

Major peace lines

Mainly Catholic

Mainly Protestant

Fire Station locations ●
circa 1969

© Mags McKay

Glossary and Abbreviations

Appliance – A generic term for a fire engine used within the service (machine is also used for this purpose.)

ATO – Ammunition Technical Officer – Mainly used for bomb disposal in Northern Ireland, although they have a wider responsibility for all aspects of ammunition within the British army.

BA – Breathing Apparatus – Self-contained device worn by firefighters, which provides breathable air in a hazardous atmosphere.

Catholic Ex-Servicemen's Association – A nationalist vigilante group, formed in the early 1970s and mainly confined to west and north Belfast.

Claymore – A generic term used to describe an anti-personnel mine.

CS gas – A tear gas used in riot control. It worked by causing a burning sensation and irritation to the eyes, nose, mouth and throat. Makes it difficult to breathe and see.

DUP – Democratic Unionist Party – Founded in 1971, it attracted a strong working class Protestant vote but in recent years has tried to widen its appeal. It is currently the largest party in the power-sharing Northern Ireland Assembly.

Ferret – A five-ton light armoured patrol car with a Browning machine gun mounted on top, the army used it during the early part of the Troubles.

H-Blocks – Compounds in the Maze Prison. Named because of their shape, they contained 200 cells each.

Height Appliance – A generic term, used within the fire service, for a hydraulic platform or turntable ladder.

HP – Hydraulic Platform – In addition to the functions of the turntable ladder this vehicle also provides a secure platform from which firefighters can operate. The operating arm can bend in one or more places and this allows more flexibility in positioning and manoeuvring the platform.

Kneecapping – A form of punishment shooting used by the paramilitaries during the Troubles. A shot was fired through the knee of the victim causing permanent damage.

Orange Order – Founded in County Armagh in 1795. By the time of the late 19th century it had expanded into a political and religious grouping, which opposed nationalism and British government efforts towards constitutional change. Its tradition of marching, sometimes through nationalist districts, can cause controversy.

Peace line – Originally large fences made from corrugated metal, the army erected them in the early 1970s to provide a physical barrier between Catholic and Protestant areas. They are located mainly in north and west Belfast and some of them have been replaced with high brick walls.

Pig – The nickname of a one-ton Humber armoured car, used throughout the first two decades of the Troubles by the army as an armoured personnel carrier.

Retained Firefighter – A firefighter who typically works at another job, and is alerted to come into their fire station when there is a fire call. Predominantly used outside larger towns and cities, this type of firefighter was employed throughout Northern Ireland, with the exception of Belfast and Derry/Londonderry, at the start of the Troubles.

Robot used by the ATO – See wheelbarrow.

RPG-7 – An anti-tank, rocket-propelled grenade, which was shoulder-fired. The IRA, and occasionally loyalist groups, used it. It had a very effective penetrating capability if it hit the target, but it was inaccurate and would self-destruct at about 1,000 metres.

RTA – Road traffic accident.

RUC – Royal Ulster Constabulary.

Provisional Irish Republican Army – Generally known as the IRA, or PIRA to the security forces. The largest of the republican paramilitary groups it was formed after a split from the official IRA.

Proxy bombs – In order to reduce the risk to themselves, bombers would often hi-jack a vehicle with two occupants, or a vehicle from an owner, hold one of the occupants of the vehicle or a member of the owner's family hostage, load a bomb onto the vehicle and order the driver or owner to drive the vehicle and bomb to a target. Many of these attacks would be less than successful since the frightened driver would often abandon the vehicle before it was fully taken to the target.

Rubber bullet – A weapon used by the security forces, said to be less lethal than gunfire. Replaced by plastic bullets in 1975 and officially called a baton round.

Sangar – A sandbagged or concrete emplacement, usually providing observation at the corner of a barracks building or a lookout position.

Saracen – A six-wheel drive, ten-ton armoured personnel carrier, used throughout the Troubles.

SDLP – Social Democratic and Labour Party – A nationalist political party, it was founded in 1970 with the aim of promoting a united Ireland by peaceful means. It was the largest nationalist party in Northern Ireland but since the 1990s it has lost considerable ground to Sinn Féin.

77 (seven-seven) – Brigade code for an explosive device, an actuated 77 was a device that had detonated.

SAS – Special Air Service – A special forces unit of the British Army.

Sinn Féin – Initially regarded as the political wing of the IRA. An all-Ireland political organisation claiming descent from a party formed in 1904 by Arthur Griffith, which took its current form in 1970. Currently it is part of the power sharing Northern Ireland Assembly where it is the second largest party.

Stormont – The building, completed in 1929, which housed the Northern Ireland parliament until it was dissolved in 1972. It became the seat of the Assembly established after the 1998 Good Friday Agreement.

TL – Turntable Ladder – A telescopic ladder, mounted on a turntable and used to gain access to fires at height. Can allow for a jet of water to be directed into the upper floors of a building or provide an external staircase for rescue purposes.

UDA – Ulster Defence Association – The largest Protestant paramilitary organisation. Established in Belfast in 1971.

Ulster Unionist Party – A unionist party in Northern Ireland that ran the state from 1920 to 1972. It was called the Official Unionist Party to distinguish it from the DUP and other splinter groups that emerged during the 1970s. Since the 1990s it has lost considerable ground to the DUP.

UVF – Ulster Volunteer Force – A loyalist paramilitary group established in the mid 1960s by Shankill Road loyalists. It carried out the first killings of the Troubles.

Ulster Workers Council – Linked to the UDA it emerged in order to organise the loyalist strike that brought down the power-sharing executive in 1974.

Wheelbarrow – A sophisticated piece of equipment used by the ATO, it is remote controlled, mounted on tank-tracks, has a weapons platform and a television camera. It can carry a variety of equipment, explosives and weapons. The main limitations were the batteries that powered it, which could run out of power, the 100m-control cable, that could get trapped or damaged, and difficulty in operating on soft ground, despite the tank-tracks.

Wholetime Firefighter – A firefighter who works full time for the fire service, typically on a watch based shift system, or nine to five in fire prevention. Employed in cities or larger towns, this type of firefighter was employed in Belfast and Derry/Londonderry during the Troubles.

Contributors

Brian Dynes: Joined the Brigade 1963, left the Brigade 1986.

Jimmy Armstrong: Joined the Brigade 1965, left the Brigade 1996.

Ken Harper: Joined the Brigade 1965, left the Brigade 1999.

Dessie McCullough: Joined the Brigade 1966, left the Brigade 1996.

Bob Pollock: Joined the Brigade 1966, left the Brigade 1998.

Roger Dawson: Joined the Brigade 1969, left the Brigade 2004.

Ken Spence: Joined the Brigade 1969, left the Brigade 2007.

Jim Hughes: Joined the Brigade 1970, left the Brigade 2000.

Stanley Spray: Joined the Brigade 1970, left the Brigade 2004.

Charlie Hughes: Joined the Brigade 1971, left the Brigade 1999.

Ken McClune: Joined the Brigade 1971, left the Brigade 2001.

Harry Welsh: Joined the Brigade 1971, left the Brigade 2003.

William Hoey: Joined the Brigade 1972, left the Brigade 1996.

Walter Mason: Joined the Brigade 1974, left the Brigade 2004.

Joe Sloan: Joined the Brigade 1974, left the Brigade 2002.

William McClay: Joined the Brigade 1975, left the Brigade 1997.

Gordon McKee: Joined the Brigade 1975, left the Brigade 2005.

Kenneth McLaughlin: Joined the Brigade 1975, left the Brigade 2010.

Murray Armstrong: Joined the Brigade 1977, left the Brigade 2009.

Wesley Currie: Joined the Brigade 1977, left the Brigade 2010.

Colin Lammy: Joined the Brigade 1977, left the Brigade 2010.

Louis Jones: Joined the Brigade 1978, left the Brigade 2010.

Gordon Latimer: Joined the Brigade 1978, left the Brigade 2009.

Gordon Galbraith: Joined the Brigade 1981, left the Brigade 2006.

Nick Allaway: Joined the Brigade 1984, left the Brigade 1994.

Brian McClintock: Joined the Brigade 1984, left the Brigade 2015.

Chris Kerr: Joined the Brigade 1985, left the Brigade 2015.

Jim Quinn: Joined the Brigade 1986, to date (2018).

Where the contributors are referenced, their words have been edited to prevent repetition and improve readability. However, great care has been taken to be true to their meaning and their actual words have been used as much as possible, to ensure that their own voices are heard. They all served in a number of Belfast stations, the exception being Jim Quinn who served in the emergency control centre. Some remained station based for all of their careers, some were promoted through the ranks and some were trade union representatives.

Chronology

1969

April — The 'troubles' come to Belfast, explosions disrupt water supplies and firefighters are kept busy with fires started deliberately.

July/August — Street rioting returns to the city as the violence and burning worsen. Firefighters are attacked with stones and petrol bombs. On the 13th of August Billy White and Dickie Sefton are badly injured when their car is petrol bombed on the Falls Road.

14th August — Soldiers are deployed on the streets of the city as the situation deteriorates.

10th September — The first 'peaceline' is constructed by the army at Cupar Street, between the Falls and Shankill Roads.

1970

April — The first major clashes between nationalists and the army occur in Ballymurphy.

13th July — One of the first 'blast incendiary devices' explodes on the Antrim Road.

1970 — The first bombs planted by the Provisional IRA in Belfast, the UVF planted a number of small devices at the homes of pro-reform politicians.

15th January — The first concerted incendiary attack in the city.

1971

9th August — Internment without trial was introduced. Thousands of soldiers were involved in early morning raids detaining three hundred and forty two people from republican areas. As a result the violence increased considerably as did the number of incidents firefighters had to attend. Subsequently, there was a spike in troubles related incidents on and around the anniversary of internment for many of the years that followed.

4th December — McGurk's Bar in North Queen Street was bombed. It was the biggest loss of life during the troubles until the Omagh bombing in 1998. The bomb detonated in the busy bar reducing the building to a pile of smouldering rubble. Firefighters working at the locus, helped by ambulance workers, the security forces and local people, pulled away the debris with their bare hands. Fifteen people died.

6th December — The fire in Munton Brothers shirt factory on the Dublin Road was described at the time as one of the biggest since the Blitz. In a collapse of the building, Salvation Army worker Mary Thompson was killed.

1972

30th January — Bloody Sunday, when fourteen people were shot and killed by paratroopers during a protest march against internment in Derry/Londonderry. Considered to be one of the most significant events of the troubles, it had enormous ramifications. Not only were fourteen lives lost, but attitudes hardened significantly and paramilitary recruitment increased. The increase in violence was immediately felt by the firefighters of Belfast as much more violence and many more deaths were generated.

4th March — The Abercorn Restaurant bomb, two young women were killed and over seventy people were injured, many of the injured were badly mutilated. Firefighters rescued the victims, several of whom had limbs blown off.

20th March — A car bomb in Lower Donegall Street. Seven men lost their lives and one hundred and fifty people were injured. Some of the bodies were carried away in fire brigade tarpaulins.

1st April — A new type of explosive, made from fertiliser, is used in a 150 lb. bomb in Church Lane. More available and easily obtained than the gelignite, which had in the main been used for bombs until then, this relatively new development would allow the production of many more, and larger, devices than would have been possible without it.

10th May — The Co-op fire, one of the largest fires of the troubles. Two firefighters had to be rescued from the roof of the burning building and it took us three days to fully extinguish it.

18th July — A major fire in J. P. Corry's where gunmen took hostage two appliances and their crews. They threatened to blow up the appliances and their crews if the Brigade didn't withdraw from the yard. They also said that any firefighters continuing to fight the fire would become targets.

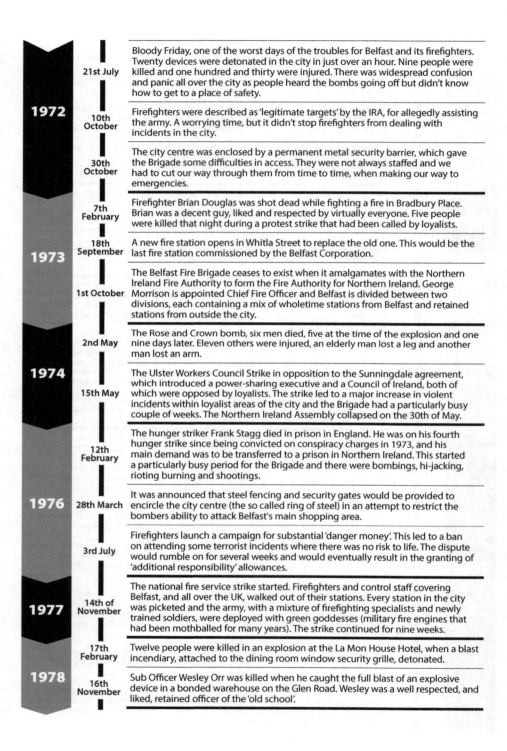

1972

21st July — Bloody Friday, one of the worst days of the troubles for Belfast and its firefighters. Twenty devices were detonated in the city in just over an hour. Nine people were killed and one hundred and thirty were injured. There was widespread confusion and panic all over the city as people heard the bombs going off but didn't know how to get to a place of safety.

10th October — Firefighters were described as 'legitimate targets' by the IRA, for allegedly assisting the army. A worrying time, but it didn't stop firefighters from dealing with incidents in the city.

30th October — The city centre was enclosed by a permanent metal security barrier, which gave the Brigade some difficulties in access. They were not always staffed and we had to cut our way through them from time to time, when making our way to emergencies.

1973

7th February — Firefighter Brian Douglas was shot dead while fighting a fire in Bradbury Place. Brian was a decent guy, liked and respected by virtually everyone. Five people were killed that night during a protest strike that had been called by loyalists.

18th September — A new fire station opens in Whitla Street to replace the old one. This would be the last fire station commissioned by the Belfast Corporation.

1st October — The Belfast Fire Brigade ceases to exist when it amalgamates with the Northern Ireland Fire Authority to form the Fire Authority for Northern Ireland. George Morrison is appointed Chief Fire Officer and Belfast is divided between two divisions, each containing a mix of wholetime stations from Belfast and retained stations from outside the city.

1974

2nd May — The Rose and Crown bomb, six men died, five at the time of the explosion and one nine days later. Eleven others were injured, an elderly man lost a leg and another man lost an arm.

15th May — The Ulster Workers Council Strike in opposition to the Sunningdale agreement, which introduced a power-sharing executive and a Council of Ireland, both of which were opposed by loyalists. The strike led to a major increase in violent incidents within loyalist areas of the city and the Brigade had a particularly busy couple of weeks. The Northern Ireland Assembly collapsed on the 30th of May.

1976

12th February — The hunger striker Frank Stagg died in prison in England. He was on his fourth hunger strike since being convicted on conspiracy charges in 1973, and his main demand was to be transferred to a prison in Northern Ireland. This started a particularly busy period for the Brigade and there were bombings, hi-jacking, rioting burning and shootings.

28th March — It was announced that steel fencing and security gates would be provided to encircle the city centre (the so called ring of steel) in an attempt to restrict the bombers ability to attack Belfast's main shopping area.

3rd July — Firefighters launch a campaign for substantial 'danger money'. This led to a ban on attending some terrorist incidents where there was no risk to life. The dispute would rumble on for several weeks and would eventually result in the granting of 'additional responsibility' allowances.

1977

14th of November — The national fire service strike started. Firefighters and control staff covering Belfast, and all over the UK, walked out of their stations. Every station in the city was picketed and the army, with a mixture of firefighting specialists and newly trained soldiers, were deployed with green goddesses (military fire engines that had been mothballed for many years). The strike continued for nine weeks.

1978

17th February — Twelve people were killed in an explosion at the La Mon House Hotel, when a blast incendiary, attached to the dining room window security grille, detonated.

16th November — Sub Officer Wesley Orr was killed when he caught the full blast of an explosive device in a bonded warehouse on the Glen Road. Wesley was a well respected, and liked, retained officer of the 'old school'.

1980	11th November	The new fire station on the Springfield Road (one of the two designed to replace Ardoyne) opened, and Ardoyne station closed. The official opening took place on 28th March 1981. Ardoyne was a special station by any measure, busy and 'right in the middle of things'.
1981	1st March	Bobby Sands, the IRA leader in the Maze prison, started a new hunger strike aimed at regaining political status for prisoners. Ten hunger strikers would die before the strike was ended in October. Following each death there was a marked increase in the violence in Northern Ireland and the Brigade's workload rose dramatically.
	5th May	Bobby Sands died. Widespread rioting broke out in the city, there was sustained petrol bombing, buses and other vehicles were burnt and used as barricades and there was a lot of shooting. Businesses were closed, the army fired hundreds of rubber bullets, and several buildings were set on fire. The Brigade was exceptionally busy.
1982	1st May	Clive Halliday, the Deputy Chief of Staffordshire, was appointed as the new Chief Fire Officer for Northern Ireland.
1984	2nd July	The Fair Employment Agency issued a report on the Brigade. They indicated that the ratio of Protestants and Catholics was unbalanced, and the number of Catholics was 'substantially below the proportion to be expected'.
	1st January	Deputy Chief Officer Billy Beggs was appointed Chief Fire Officer, following Clive Halliday's appointment as Firemaster of Strathclyde Fire Brigade.
	29th July	A 500lb van bomb totally devastated Central Fire Station.
1985	15th November	The Anglo-Irish Agreement was signed, bringing permanent inter-governmental conference machinery and a consultative role for the Irish government in Northern Ireland affairs; a joint secretariat was set up at Maryfield outside Belfast. Unionist opposition to the agreement was so deep it lasted for years and resulted in political boycotts, mass rallies and an increase in violence throughout the city. Once again this led to a major increase in incidents for firefighters, and the Brigade was particularly busy.
1986	19th July	Firefighter Martin Duffy was shot dead while working as a part time taxi driver. Martin was twenty-eight, came from Manor Street and had been in the Brigade for eight years.
1988		At the end of the year Ken McNeil, who had been in the service since 1959, was appointed as the new Chief Fire Officer for Northern Ireland on the retirement of Billy Beggs.
1991	3rd June	Due to a major reorganisation of the Northern Ireland Fire Brigade, Belfast, which had been split between two divisions as part of the amalgamation of the two Brigades in 1973, was recombined into a new city division with seven stations, as it now included Glengormley. Over the reporting year the new 'A' Division carried out 31% of the operational workload of the Brigade.
1992	27th November	Central Fire Station in Chichester Street closed. Built in 1894 the station was originally the headquarters of the old Belfast Fire Brigade, and it was the location where generations of Belfast firefighters had been trained. It was replaced by a new purpose built station and divisional headquarters in Bankmore Street. The official opening of Bankmore Street took place on the 10th of May 1993.
1993	23rd October	An explosive device demolished Frizzell's fish shop on the Shankill Road. Ten people were killed, including one of the men who had planted the device. Two children and four women were among the dead and fifty-seven people were injured, many of them seriously.
1994	31st August	The Provisional IRA declared 'a complete cessation of military operations', which was followed a month later by a Loyalist cease-fire.

Introduction

EVEN THOUGH BELFAST WAS not always a sectarian city, the roots of sectarian violence can be traced back to the early 19th century when the city was industrialising and growing rapidly. One early example of sectarian violence occurred on the evening of 12 July 1813 when some local people attacked an Orange procession attempting to march down Royal Avenue. Some of the Orangemen involved went to get muskets and shot dead two bystanders, who happened to be Protestants. Four Orangemen were convicted of murder and four Catholics were found guilty of riot. However, the wider implication was that the bitter ethnic rivalries, which at the time existed in parts of rural Ulster, were brought into town.[1] Street rioting, following Orange Order marches, continued through the 1800s and Catholics and Protestants started to segregate into separate areas. Driven by rumour, counter-rumour and violence on both sides, waves of savage disturbances occurred periodically throughout the rest of the 1800s and well into the 1900s, with an upsurge in violence in 1921, due to the division of Ireland into two jurisdictions. There are relatively few records of Fire Brigade involvement in dealing with disturbances during this time but it is known that, following the violence, which occurred in 1864, the Belfast Fire Brigade were paid £12 for extinguishing fires that were started by rioters.[2]

During the Second World War, Belfast was a target for German bombing and the city suffered badly during the Blitz.[3] The tragedy was that most of the bombs fell not on strategic targets, but on the densely populated residential streets from which very few people had been evacuated. In all, nearly half the housing stock in the city was affected, and thousands were left homeless. However, the fire service gained valuable experience, which would serve it well in the years to come.

While there was some sectarian violence between 1935 and 1962, 1963 saw a rise in street violence in Belfast in what came to be called 'the Battle

---

[1] Hepburn (1996).

[2] Boyd (1969).

[3] Maguire (2009).

of Divis Street', when the Royal Ulster Constabulary removed the Irish Tricolour from the headquarters of the Republican Party. In 1966, the Ulster Volunteer Force was formed on the Shankill Road, taking its name from a pre-1914 unionist private army raised against home rule, and, between 1967 and 1969, politics returned to the streets during the rise of the civil rights movement and loyalist reaction to it.[4]

Despite, or maybe because of, their shared Troubles, the people of Belfast are great, resilient and friendly. They can also be 'candid, very frank and not afraid to give you advice or encouragement, positive or negative'.[5] The Brigade's firefighters are an integral part of the city and share many of the characteristics of its people. All of their resourcefulness, experience and competence were required as the Troubles developed in 1969 and continued for the 25 years or more that followed. They were required to face vicious rioting, bombing, burning and killing, dealing with Troubles-related incidents on an almost daily basis; and often several times a day, unarmed and with very basic protective clothing. Throughout this time, they delivered a fully competent emergency service to all sections of Northern Ireland society, without fear or favour and at great personal risk. They lived and worked in Belfast, many of them in areas that saw the Troubles first hand, but they left their personal beliefs and loyalties behind when they came into the fire station and looked after each other and the people of their city. The firefighters responded to every incident during the Troubles, wherever it was located, seeing the best and worst of humanity, and in this way they provided an almost unique example of courage and compassion in very difficult circumstances.

Even though there is not enough room to describe every incident, or tell the story of every firefighter, I hope that the examples included here will give a strong flavour of 'how it was' for those who were there at the time. Extraordinary things were achieved by ordinary firefighters and I hope this book does them justice.

---

[4] Boyd (1969).

[5] Chris Kerr.

PART ONE

1969 to 1971

The Descent into Chaos

1969

NINETEEN-SIXTY-NINE is the year generally regarded as the beginning of what has come to be known as the Troubles, witnessing as it did the first deaths since 1966 and the arrival of the army on the streets.[6] The first quarter of the year reflected Northern Ireland's occasionally violent and sectarian history, with hints of the death and destruction yet to come. 1 January 1969 saw the start of a three-day march, organised by the Peoples Democracy, from Belfast to Londonderry in support of civil rights for Catholics in Northern Ireland. Although the march began peacefully, counter demonstrations were organised in Protestant areas and, despite a police presence, fights soon broke out when stones were thrown at the marchers. Similar scenes were seen in Newry where another civil rights march ended in violence. However, for the firefighters of Belfast, things carried on more or less as normal for a city-based emergency service.

The Belfast Fire Brigade at the Beginning of 1969

At the start of 1969 the Brigade had a total establishment of approximately 300 personnel, with firefighters based in five stations: Central (which was also the Brigade Headquarters), Ardoyne, Whitla Street, Knock and Cadogan. They operated a 'three, three and three shift' system with three day shifts of nine hours on duty, followed by three night shifts of 15 hours, followed by three days off. Therefore, there would be approximately 65 firefighters, with ten pumping appliances, three height appliances, a foam tender and

---

[6] McKittrick, Kelters, Feeney and Thornton (1999).

an emergency tender, on duty at any given time. The firefighters' personal protective equipment was pretty basic at that time. It comprised of: a cork helmet to protect the head; a woollen fire tunic with a leather belt, to which was attached a belt line (a short, narrow diameter rope used for various purposes) and a firefighter's axe; a pair of black rubberised leggings, one for each leg, they came up in a point at the front where they were attached to the belt of the trousers by a button and dog clip (their shape meant that your backside was always getting wet); and rubber boots with steel toe caps and soles.

The Brigade had a fully developed organisational culture, similar to that of many other Brigades in Great Britain. It was traditional, hierarchical and autocratic in its nature, with a militaristic management style that could be seen as petty at times:[7]

> Your axe had to be polished and your belt had to be polished, your leggings had to be polished. It was really very regimental in those days.[8]

> You were terrified of the Station Officer and there were certain things that had to be done, like floors washed and stairs washed. The brasses out the front had to be done as well; if you hadn't cleaned them properly you had to stay after nine to make sure they were cleaned properly. They had all sorts of devious things. They used to put chalk marks under the mudguards to see if the fire engine had been cleaned during the day.[9]

The Brigade was an integral part of the local community and provided a well-regarded service to it:

> It was fantastic, you had to do everything. You had to clean out sewers in the Markets. You were the doctor's surgery for the Markets. 'You may come round me Da's beating me Mother', and the Land Rover had to go round and sort things out. It was just the place they would come if they were in trouble at all.[10]

---

[7] Allaway (2011).
[8] Dessie McCullough.
[9] Brian Dynes.
[10] Jimmy Armstrong.

Whitla was like a first aid post on a Saturday night for all the drunks coming out from the dock area, they didn't go to the doctor they went to Whitla Fire Station. These two guys arrived at the door and this wee man was bent double. 'My mate's broke his back!' The two of them were drunk as skunks. So they sort of shuffled him in: 'what's wrong?' 'I can't straighten up! I can't straighten up!' So we took off his coat and everybody was falling about laughing. He had braces on and he had buttoned the braces to his spur, to the buttons on his trousers. They were holding him down.[11]

Even though the workload was similar to other large cities, the Brigade in 1969 was relatively quiet in comparison to what was to come:

It was quiet in the sense that it was house fires, normal type fires, but you were busy enough.[12]

This relatively peaceful existence started to change in April, when a series of explosions hit the water supplies of the city, causing days of water shortages. On 4 April, the Dunadry water installation was hit, on 20 April the Silent Valley water pipeline, on 24 April the water pipeline from Lough Neagh and on 25 April there was an explosion in Annalong.

The burning had started. On 7 April four fire engines dealt with a fire started by a petrol bomb in Queen's University. The fire wasn't particularly big or difficult to deal with and it was under control within 15 minutes. A more serious fire was started on 9 April in Donaghy and Kelly's, a three-storey upholstery warehouse in Corporation Street. This fire was much more difficult to extinguish and it took 35 firefighters three hours to bring under control. During the night of 20 April, at about 11.00pm, ten post offices and a bus depot on the Falls Road were petrol bombed in a seemingly co-ordinated attack. Two of the post offices were completely burnt out, one was badly damaged, and the rest, together with the bus depot, were only slightly damaged.

On 28 May there was a major fire in the Grand Central Hotel. Guests climbed onto window ledges to escape from the blaze and a turntable ladder

---

[11] Dessie McCullough.
[12] Bob Pollock.

was used to carry out rescues from the upper floors. Eighteen guests were taken to hospital for medical attention. On 5 June fire destroyed the Wimpey Bar in Wellington Place. During the course of the incident two firefighters were blown out of the front of the building when the fire flashed over and burst the plate glass window at the front of the restaurant; a third firefighter was taken to hospital with cuts to his hands.

The City Erupts

In the middle of July 1969, street rioting returned to parts of the city.[13] Stones and petrol bombs were thrown, with at least one car and several buildings damaged. The violence worsened at the beginning of August and the Brigade was often caught in the middle of riots after being called into action. On one Sunday alone, 19 fire calls were received from the Shankill/ Crumlin Road area. A number of public houses and off-licences were set on fire and petrol bombs were thrown into the Holy Cross School. Cars and vans were destroyed by fire and one firefighter was injured. Things became even more dangerous when hostile crowds set a number of houses on fire and, in Hooker Street, firefighters were petrol-bombed by rioters when they were called to the scene:

> I think once the Troubles started it just hit us with a bang, because they were burning everything. We were leaving the station at maybe half six at night and never got back till after nine in the morning; you were just going from one call to the other. They just burnt everything in sight, they really did.[14]

Dickie Sefton and Billy White

After a few days of relative calm, the violence and destruction started again on the night of 13 August. The trouble started in the west of the city when a crowd of about 200 people attacked Springfield Road Police Station, which was in the process of being rebuilt at the time. Arming themselves with sticks, stones and bricks from the building site the crowd marched down

---

[13] Elliot and Flackes (1999).
[14] Dessie McCullough.

the Falls Road, breaking shop windows as they went. The police dispersed them into side streets, but they soon reformed at Divis Street where they attacked the Hastings Street Police Station with petrol bombs. During this attack a young girl was enveloped in flames when her clothing caught fire. Burning barricades were formed and burning tyres were used to set fire to two factories in Northumberland Street. Petrol bombs were thrown into a car showroom and a number of cars were taken and used to form a burning barricade, which blocked the Falls Road.

In response to this event, pairs of Brigade officers were given the job of assessing the various situations happening across the city in order to prioritise the workload of the Brigade and help in mobilisation decisions. As the Falls Road team made their way down the Falls Road, two petrol bombs were thrown into their car, the interior of which was immediately engulfed in fire, and the two men were immersed in the flames. Assistant Divisional Officer Richard (Dickie) Sefton lost control of the car and it careered from one side of the road to the other and hit a pedestrian, eventually coming to a stop, still on fire, at the side of the road against the gable end of a house in Peel Street.

Richard Sefton, on recalling the incident, said:

> That was really the first big night of the rioting and the Assistant Chief Officer, Billy White, and I went out to survey the situation. We drove down the Falls Road, the rioting crowd surrounded us, then smashed our windscreen and chucked in two petrol bombs. Consequently the car went out of control and I remember knocking down somebody but I couldn't see, and then I lost control completely and mounted the footpath, eventually crashed into the gable of Peel Street where the doors, lucky enough, burst open. Now if we had been wearing seat belts we would have been burnt to death. Billy White got out, and I rolled out and tried to roll (the fire on) myself out. At this time I couldn't get oxygen because the flames were starting to suffocate me. I gathered later that some old lady from Peel Street ran over and beat me out with her bare hands, and then ran off. While I was lying there I heard this voice saying, 'That's not a policeman that's a fireman' and another voice said, 'Let the bastard burn'. After that the ambulance duly arrived and I was taken to hospital, and then,

for three days, I didn't know whether I was blind because my eyes were fused shut, and then my hands were stuck together. So after that it took a period of two years with all the operations, 15 operations.[15]

At the time, this incident was seen as one of the worst cases to come out of the rioting in the city and the injuries to both men were horrific. They had suffered permanent and severe scarring to their faces, ears and particularly their necks and hands, every exposed part of their bodies had been badly burnt. Both men endured their terrific pain with fortitude and courage. Dickie had suffered horrendous damage to his appearance, but the worst damage was to his hands, both of which were permanently disfigured. Even though the effects of the burning were permanent, both men returned to duty in the fire service. During this incident, though they had been advised by the police to stay out of the area, the Brigade continued to provide a service to all the residents of the city. A local Member of Parliament, Paddy Devlin, paid tribute to the Brigade saying that, 'firemen showed real courage in front of a violent mob'.[16]

* *

On 14 August troops arrived on the streets of Belfast. I lived just off the Upper Crumlin Road at that time and, together with two of my friends, decided to take a walk down the Crumlin Road to see what was going on. Soldiers were lined up in the middle of the road, more or less on the white line, about a metre apart and facing in alternate directions. As we walked past an army Ferret armoured car, which was also in the middle of the road, its turret-mounted gun followed us down the street. It was a strange, somewhat surreal experience, but there was nobody apart from the soldiers about and somehow the area seemed quite calm.

Unfortunately the arrival of the army on the streets did not stop what was by then being called 'The Troubles', and that night was a night of riots and arson, with over 30 major fires and hundreds of petrol bombs being thrown by rival mobs.

There was more trouble on the 15 August when barricades of lorries and buses were erected and set on fire. By dawn of 16 August, firefighters had

[15] Richard Sefton quoted in Forgotten Heroes.
[16] *Belfast Telegraph*.

dealt with hundreds of fires and the smoke had gathered over the city in one massive pall. There was a palpable smell of smoke and burnt timber in the air and several large fires were still burning. A Fire Brigade spokesman said of the night that 'more than 100 houses had been on fire and, in Bombay Street alone, 33 house were blazing. The fireman have been working nonstop throughout the night'.[17]

> In Bombay Street, there were 'only two jets in use, one at the end of each street. [By the time the firefighters] reached the middle of the street, the houses were falling down'.[18]

Having finished school that June, I was looking for a job. I watched the smoke, which engulfed the city from a vantage point on the Crumlin Road, and in response to an advert in the *Belfast Telegraph* that read:

> Left school this year? Don't settle for an ordinary job, choose a career with an exciting future! Be a JUNIOR FIREMAN in the Belfast Fire Brigade. Age limit – Over 16, under 17 on 1 August 1969. Pay – £340 p.a. at 17, £765 p.a. at 18. Free Uniform – Promotion Prospects – Day Release. Four weeks holiday. Closing Date – Monday, 25 August, 1969.

I decided to apply for a job with the Belfast Fire Brigade, a decision which would shape the rest of my life and one that I would never regret. Station Officer Paddy Heggarty was in charge of the training centre when I joined the Brigade; he was a jovial character and his son Joe would later be on the same watch as me in Chichester Street. Paddy was an amusing guy who used stories to illustrate his message when training recruits. He used one, told in a humorous manner but nonetheless giving a message, to let us junior firefighters know where we fitted in the hierarchy of the organisation:

> Let me tell you where you fit into the Brigade. If I come in in a bad mood one morning I'll shout at the Sub Officer, he'll shout at the Leading Fireman who'll shout at a fireman. The fireman will kick

---

[17] *Belfast Telegraph*.
[18] Dessie McCullough.

the station dog, which will bite the station cat and the station cat will scratch you, and that's where you fit in.

But it was Leading Firefighter Ronnie (Dinger) Bell who with humour, patience and discipline, when necessary, seemed to know how to keep a bunch of 16-year-olds interested in, and committed to, the Fire Brigade. And it would be Ronnie who, for the next two years, would be responsible for turning us raw recruits into firefighters.

At the same time others were making similar decisions to mine, some of them in much more difficult circumstances:

> In August 1969, our house was burnt down, the Fire Brigade were involved but had to evacuate. A street called Bombay Street basically disappeared. There was large scale rioting, the firefighters were coming under immense pressure, they were evacuated in the back of an ambulance out of the area and they had to leave their appliance. Stoned, petrol bombed, the whole heap, there was actually gunfire at times. Anybody who was in uniform, anyone who looked as though they came from authority, were targets. My father, who had been a firefighter, my two brothers, a couple of other people and I used the fire engine that was left to try and put out the flames. Got a lot of publicity, so when I tried to join the Fire Brigade, which I had done prior to this, I certainly found it slightly easier to get in on that second attempt than the first attempt.[19]

Although the death, damage and destruction didn't rise to the levels seen in August of the year, the last four months of 1969 continued in more or less the same vein. One incident of particular note seemed to bode ill for the city when, on 21 September at about 6.30pm, a bomb, made from gelignite with a slow burning fuse and wrapped in brown paper, exploded in the docks area. On 5 November there was an explosion at the Ormeau Road gas works near the city centre when a small charge of gelignite started a fire in a gasholder. Despite the obvious danger, 30 firefighters fought the blaze using jets of water; they managed to extinguish it in about 30 minutes. It was a 'dangerous' incident, brought under control by setting up a 'water curtain

---

[19] Jim Hughes.

between the blazing gasholder and another curtain about 120 ft. away to stop the fire spreading'.[20] This all happened alongside the 'normal' incidents of fire that were happening continuously across the city.

Peace line

The rioting in Belfast and the burning of hundreds of houses in the city left thousands of people, mainly but not exclusively Catholics, homeless. By 18 August, 400 refugees were accommodated in army camps south of the border. By the end of the month 1800 families had moved homes as a result of the disturbances.[21] In an attempt to provide at least some protection, barricades were erected by local people on both sides of Cupar Street, between the Falls and Shankill Roads, as communities continued to segregate. Consequently, when the army completed a 'peace line' on 10 September[22] by erecting iron sheets as a barrier, it was a physical recognition of a division that already existed. Over the following years peace line between Protestant and Catholic areas were expanded and their construction became more sophisticated and permanent. By 2007 there were approximately 40 peace walls in Belfast, stretching nearly 13 miles in total.[23]

In an early recognition of the realities of the situation, and in an attempt to help its members gain access to the areas worst affected by the unrest, the Brigade delivered hundreds of leaflets to the effected areas at the beginning of September 1969:

> **_Your barricades have been erected to protect you: help us to ensure that they do not kill you._**
>
> Fire is still the greatest enemy of life and property and while your barricades may allay your fears, they will hinder the Fire Brigade from giving you assistance. Do not forget the lessons of the past; these show that during the year there are likely to be 500 houses burned, 1,000 chimney fires, 50 persons severely burned and five deaths from fires. Belfast Fire Brigade has been helping you since

---

[20] *Belfast Telegraph.*

[21] McKittrick, Kelters, Feeney and Thornton (1999).

[22] Elliot and Flackes (1999).

[23] Gillespie (2010).

1845. It exists to save life and to protect property. It turns out 12-15 seconds after a call is received. This can be wasted if we cannot get through to help. Two minutes may make a small fire a big fire and may be the difference between life and death. Help us to help you. Please make your barricades such that you can move them quickly when the emergency arises – it may be your child, or your wife, or your home that is in danger.

1970

FOR THE FIRE Brigade in Belfast in 1970 the normal work continued, much as it had in 1969. There were the usual fires in the home and loss of life to contend with, for example, at about 10.00am on the morning of 9 February, a team of workers were laying water pipes in a ten-foot deep trench at Stranmillis College, when suddenly the trench collapsed. The two men who were working in the trench at the time tried to escape by running out of the danger area. Unfortunately they were unsuccessful and were both trapped under tons of earth, at opposite ends of the trench. Their workmates tried frantically to rescue them. Firefighters arrived within minutes and joined the rescue efforts but to no avail. Workmen stood around in shock while firefighters removed pieces of wooden shoring and tons of clay to get to the trapped men. Both men were given first aid at the scene but they were pronounced 'dead on arrival' at hospital. On 18 March firefighters carried out an animal rescue when a donkey was rescued from deep mud at the side of the Holywood Road. The next evening there was a fire in three bitumen storage tanks at the Belfast Gasworks. This fire started as workers were breaking up the 500 gallon tanks and a spark caused the bitumen sludge in the bottom of one of the tanks to ignite. This kind of fire is particularly difficult, and often dangerous, to deal with and the specialist foam tender was used to extinguish it with a covering of foam.

There were also a number of more unusual incidents to be dealt with and, in February, a soldier received slight burns to his face when an oil stove was accidentally overturned setting fire to his observation post. Several rounds of ammunition detonated but thankfully no one was injured. In May, during a particularly violent storm, a fire that was started by lightning damaged two semi-detached houses.

Despite the obvious challenges these incidents brought to firefighters, they were overshadowed by the necessity of delivering a service in what was rapidly becoming a war zone. Although the future could not be predicted at the time, 1970 foreshadowed all of the additional difficulties and challenges that firefighters would have to face over the following decades of the Troubles. Rioting, shooting, burning caused by rioters, petrol bombs, firebombs, incendiary devices, and bombings. All of these factors would increase and worsen in the years to come.

The Rioting and Shooting Intensify

When the army arrived in Belfast during August 1969, Catholics had generally welcomed them as a protection against loyalist attacks. However, this situation changed over the following months, and, in April 1970, Ballymurphy saw the first major clashes between nationalists and the army.[24] The violence continued, and, over the second weekend in May, the rioting moved to the New Lodge Road with five hours of bitter street-fighting on the Saturday night. As the army kept rival mobs apart, fires were started in several areas of the city with petrol bombs, keeping the Brigade very busy. The trouble continued until daybreak.

On the Sunday night of the following weekend the army fought continuous battles with street mobs in two areas of the city – Ardoyne and the New Lodge. The disorder began after three petrol bombs were thrown through the front window of a garage at the junction of Alliance Avenue and Alliance Road in Ardoyne. The office of the garage was soon well alight but, as the Brigade were dealing with the incident, the army and police arrived. This seemed to incense the crowd that had gathered and very soon rioters were throwing bricks, stones and petrol bombs, the army responding with cs gas as the Fire Brigade tried to extinguish the fire. Meanwhile three storeys of a spinning mill were burning in Flax Street and the Brigade required army protection from a crowd of youths who stopped them from attending the incident.

The shooting and rioting made things very difficult for the Brigade as they dealt with the various fires, but even in the most difficult circumstances the characters involved could keep life interesting:

> I remember many, many occasions where we had to stand behind barricades where the police and the army were... Churchill Street, round Hooker Street, all round the whole Ardoyne area, Twaddell

---

[24] Gillespie (2008).

Avenue, stretching right up and across the West Circular... that whole segment had street violence almost on a daily basis. There was a lady known as Skinny Lizzie who was the only Protestant in Hooker Street and she decided she would fly the Union Jack on the 12th. They burnt Skinny Lizzie out eight times and we rescued her twice, resuscitated her once and eventually the poor old girl in her late '70s died. She was a big woman over six foot and she didn't take any nonsense from anybody. [25]

By the beginning of July a second peace line had been built on the Crumlin Road to prevent violence in the Hooker Street, Disraeli Street area. Over the first weekend of August, soldiers fought running battles with rioters in Ardoyne, Andersonstown, the Falls Road and the New Lodge, and shots were fired at soldiers in the Whiterock area. Trouble also flared at Ballymurphy and Divis Flats. On the sixth night of rioting, the rent offices of the Northern Ireland Housing Trust were seriously damaged by fire following a petrol bomb attack. Two banks and nearby shops were also attacked with petrol bombs. Firefighters extinguished the fires at the banks and shops as the disturbances continued, but despite their attempts to contain the fire at the housing trust the building was badly damaged. There was also a certain amount of stupidity, brought about by drink. On 26 August, the Brigade was called to a house fire, which was caused by a petrol bomb. The three men in the house were drunk and had been trying out a petrol bomb to see if it would work; it did. Thankfully, even though there were six children in the house at the time, no one was seriously hurt. The rioting continued sporadically throughout the rest of the year with lorries and vans being hijacked and used as barricades and often being set on fire.

Burning

1970 saw the initial use of explosive incendiary devices in the city and on 13 July an explosion badly damaged the foyer and ground floor of a hotel on the Antrim Road. A fire followed the explosion and it was believed that an incendiary device had been used. A fire in the Old Lyceum Cinema, which was used to store thousands of feature films for the Rank Organisation, illustrates the difficulties faced by firefighters at this time. Nearly three thousand films, about five years of Ulster cinema entertainment, were

---

[25] Ken Harper.

destroyed during rioting on 31 July. The Brigade was called and attended but initially were unable to attack the fire for some time, due to the rioting. By the time firefighters were able to gain access to the building it was well alight. The fire was brought under control using six jets of water but, largely due to the delay in getting to work, the building was severely damaged and all of the films were destroyed.

Bombs

The year saw the first bombs planted by the Provisional IRA, and the UVF planted a number of small devices at the homes of pro-reform politicians.[26] The early bombs were quite primitive: they used timers comprising clothes pegs held open by rubber bands with tintacks as contacts. The spring in the clothes peg gradually overcame the strength of the rubber band and when the tintacks came together the device would detonate.[27] When these devices were pre-prepared they would often have a piece of dowelling inserted in the clothes peg to keep the contacts apart and the device was primed by removing the dowelling.

The Role of the Brigade at Explosions

The role of the Fire Brigade had been well defined for generations of firefighters before the Troubles started. It was to 'save life, protect property and to carry out such humanitarian duties as may be necessary'. These underlying requirements of the job were taught to all fire service recruits and were well understood and internalised within the culture of the service. There was a statutory requirement to fight fires contained in legislation, but bombs were something different. At normal air pressure and normal temperature the pressure inside and outside a person's body is 14.7 lbs per square inch. An explosion causes a wave of energy to pass over a person and if the pressure increases by more than five lbs the lungs collapse, limbs can be torn off and the person dies. If the detonation is in an enclosed space, such as a room in a building, the damage to people and the building are magnified.[28]

---

[26] McKittrick, Kelters, Feeney and Thornton (1999).

[27] Styles and Perrin (1975).

[28] Styles and Perrin (1975).

Nonetheless the role of the Brigade at bombing incidents was never questioned by its firefighters. We fully accepted that we would attend to save life where we could, to protect and make safe property and to do what was necessary to carry out our humanitarian duty. If a fire was involved the senior fire officer present would take command. Where there was no fire the police had overall control, while firefighters worked to carry out rescues, made the buildings or vehicles involved as safe as possible and turned off the gas and electricity or other potential sources of ignition. Often, our main task was body recovery.

A number of devices exploded during 1970. To take just one example, on 26 February an explosion badly damaged the Belfast Corporation Electricity Club in Upper Donegal Street. The 25-year-old barman was blown through the doorway:

> There was a sudden thud and a ringing in my ear and I was blown through the door into the club. As I lay there I could see the bar moving sideways and bottles and crates falling on top of my colleagues. At first I thought it was a beer barrel or a gas cylinder exploding then someone shouted, 'It's a bomb! It's a bomb'.[29]

In July the bombing escalated and between Friday 3 July and Saturday 4 July, bombs damaged six buildings. On 16 July, 24 people were injured, three of them seriously, when a bomb detonated at the Northern Bank in the High Street. Glass panelling and brickwork were blown out of the front of the bank and broken glass and flying debris injured both passers-by and customers. Three fire engines attended to assist with the casualties and make the building safe. The bombing continued for the rest of the year.

1971

NINETEEN-SEVENTY-ONE SAW A MAJOR escalation of violence, and the death toll rose sharply following the introduction of internment in August. Amid the continuing political upheaval the rioting and bombing got worse, and gunfights between the army and the IRA, often associated with major civil disorder, became commonplace. Rioting continued in several areas of the city, with bottles, stones, petrol bombs and nail bombs being thrown. Even more worryingly, the rioters used machine guns, bottles of acid, even

---

[29] *Belfast Telegraph.*

steel tipped arrows from a bow, and on several occasions firefighters were attacked when attending fires in difficult parts of the city.

The Rise of the Incendiary Device

On the night of 15 January the first concerted incendiary attack was carried out in the centre of Belfast.

The burning started at about 5.15pm when staff at Robb's in Castle Place found a device under a cushion in the shop. The small fire it started was quickly extinguished and there was little damage done. An hour or so later a fire broke out in a furniture shop in York Street. This was a more significant fire and firefighters struggled to bring it under control. Showrooms and offices were destroyed as the fire damaged the upper floors and roof of the building. At 7pm, Boots the Chemist in Cornmarket was on fire. This was a slightly easier fire to deal with, due to a lesser fire loading, and firefighters restricted it to the upper floor, preventing serious damage to the rest of the building. Fifteen minutes later a fire was started by an incendiary device hidden in one of the pockets of clothing for sale in Fraser's outfitters, but only the clothing itself was damaged. At about 7.45pm, as firefighters were still dealing with the fire in Boots, they noticed smoke issuing from the furniture store of Hanna and Brown in Arthur Street. This was the most difficult fire of the night and firefighters needed to wear breathing apparatus to enter the building in order to fight the fire at close quarters. The shop and its contents were badly damaged before the fire could be extinguished. At about midnight an incendiary device was spotted in a letterbox in Talbot Street. Although the device had ignited there was no fire spread and little damage was done.

At about the same time there was a small fire in the Gibraltar Bar in York Street. Firefighters were quickly on the scene and extinguished it. At about 3.00am on 16 January, a major fire was discovered in the Star and Garter Bar in Rosemary Street. Firefighters used five jets of water to contain the blaze but the three-storey building was extensively damaged before the fire could be extinguished. The Brigade was still in attendance, damping down the debris and rubble, several hours later. As firefighters fought to control the spate of incidents the key-holders of shops and offices in the city were asked to return to their premises and check them for devices, as firefighters found several undetonated 'clothes-peg' incendiary devices at a number of the incidents they were dealing with.

On 20 January a number of incendiary devices were planted in Wilson Brothers Sack Merchants in Hill Street. Firefighters quickly put out a relatively small fire in the 12 bales of sacks involved and there was no damage to the building. A clothes-peg incendiary device, which hadn't detonated, was found in the building and it was believed that a number of similar devices started the fire. Another incendiary device was found at 5.00am on the 22nd in a bus at Smithfield Bus Station; it did not ignite and was disposed of by putting it into a bucket of water. As the Brigade was attending this incident a fire engine was stoned while attending a malicious call in Andersonstown. It was thought that the call had been made deliberately to try and tie up fire engines as incendiaries in the city detonated. All in all the first month of the year had been so busy that a local paper reported on the 27th that 'Belfast had a relatively quiet night, the only incidents being two small blasts and the discovery of two incendiary devices'.[30]

Late at night on 9 February firefighters had to deal with the biggest fire of the Troubles yet when a massive fire swept through a timber yard belonging to McCue, Dick and Company in the docks. The fire destroyed a warehouse, and half of the timber in the yard. At one point just about every firefighter available in the city fought the fire, under the command of the Chief Fire Officer, Robert Mitchell. It took ten hours for firefighters to bring the fire under control and they were still in attendance over 24 hours later. The value of the material and building destroyed was said to be worth around one and a half million pounds.

Dessie McCullough remembers the incident:

> The whole shed was just ablaze. So we got into it anyway and I think there were about four jets on it and all you heard was this boom. I said, 'what was that?' and on down there was another big bang, so there were incendiaries planted all over it. You know McCue Dick's was like a quarter of a mile, a massive yard. So they sent this squad from the police, I don't know whether they were recruits or who they were, but those guys moved, they made a fire break. They must have moved 100 ton of timber and the sweat was lashing off them. So they made this fire break about maybe ten yards wide and we said now if we get a jet on it'll not go any further than that. So we were running round with jets and the next thing boom! On the other side of it (the fire break) and away it went, 33 jets on that fire.[31]

---

[30] *Belfast Telegraph.*
[31] Dessie McCullough

On 16 April there were more than a dozen incidents in the city, the worst of which involved a number of offices and shops in Lower North Street, where it took more than 30 firefighters to control a fire. A Fire Brigade spokesperson said: 'This was the most serious incident in a very busy night for us'.[32] The burning continued, and, in May, attacks using incendiary devices reached a peak. On the second weekend of the month, 42 devices were left in city centre stores, 36 of them on Saturday 8 May. It was only swift action by firefighters and the security forces that prevented massive destruction that night. The devices, contained in cigarette packets and hidden in the various shops, were only discovered when over a dozen of them ignited. The timing of those planting the devices was probably misjudged as the first devices activated in Anderson and McCauley's in Donegall Place before closing time, allowing for the early detection of many other devices. In all 16 devices were found in Anderson and McCauley's but they were disposed of and little damage was caused. Shortly before 6.00pm, three devices ignited in the Belfast Co-op in York Street and an additional 12 were discovered in a subsequent search of the premises. A further six ignited in the North Street branch of Woolworth's and two others were found there. One of two devices planted in a card rack in Candles and Cards in Wellington Place ignited, and two more were pushed through the letterboxes of other shops, one in Babyneeds in Donegall Square West and the other in May's House Furnishers in May Street. At the same time, just to keep everybody on their toes, a small bomb containing four ounces of gelignite, damaged the frontage of a grocer's shop in Madrid Street:

> These incendiary devices were very small and could be slipped into the pocket of, say, a coat hanging on a rail in a department store. The favourite ploy was to place them throughout a store just before it closed its doors for the day. When all the staff had left the devices would start igniting and set the store ablaze. It was extremely difficult for the terrorist to get his timing right, but usually the incendiaries ignited between 20 and 30 minutes of being placed. One Saturday afternoon, early in May, the terrorists had laid 36 of these devices at busy stores in the centre of Belfast. Our EOD rushed in after the first one was discovered and managed to find 18 incendiaries before they ignited. These were lifted carefully from their hiding places and dumped in buckets of water. Warrant Officer 'Noddy' Park went back to check the scene after the first panic had died down. An RUC officer walked up to him

---

[32] *Belfast Telegraph.*

outside a store. 'We've found another incendiary that hasn't gone off sir,' he said proudly. 'We've put it in the back of that police Land-Rover there if you want to have a look.' 'Noddy' Park did have a look. There, on the back seat just behind the driver, was the device. 'Get out of the way, for Christ's sake,' he shouted at the policemen. Then he leaned inside the Land Rover, picked up the incendiary device with his bare hands, ran a few steps and threw it down the road away from everyone. It exploded before it hit the ground but no one was near enough to be hurt. A few seconds later it would have exploded behind that police driver. The effect would have been similar to playing a blowtorch on the back of his head, slowly. In another word – horrible.[33]

Firefighters were not immune to this kind of naiveté:

> A lot of incendiary bombs, they used to go off in front of you. We used to go in and we didn't know any different, just going in and putting them out and they were going off beside us. We didn't know any better, we thought it was part of our job, you know, looking for these things.[34]

> Roy Lemon and I were on breathing apparatus. In behind the row of houses, there was a wee workshop. This was smoky anyway, and we all had lectures in the stations about incendiary devices, which were made out of Embassy cigarette packets with a piece of dowelling wood shoved through. So Roy and I found this pile of timber on fire that was causing all the smoke, knocked it out with a hosereel. And I looked round and saw one of these cigarette packets, exactly the same as had been in our lecture the week before. So I called Roy over and pointed it out to him. Roy then came over with a steel wastepaper bin and I lifted the incendiary device and set it into the steel bin and we walked down. Anyway we found eight of these, the ATO had arrived and we very proudly walked out and presented him with a steel bin with eight incendiary devices in it. He ran outside, up the entry and evacuated the area; we had made a bomb.[35]

---

[33] Styles and Perrin (1975).

[34] Jimmy Armstrong.

[35] Ken Harper.

On 9 May, the 67-year-old mother of the leader of the Red Hand Commando and one-time chairman of the Shankill Defence League, died in a petrol bomb attack at the family shop on the Albertbridge Road. The petrol bomb was thrown into her son's shop below their flat. Two men were treated for burns after trying to rescue her.[36] This was one of the most difficult incidents firefighters had to deal with that weekend, not only because of the fatality but also because of the tension it caused in the area. On Saturday 19 June several incendiary devices detonated and damaged shops in the city centre. The most serious damage was caused to Millets in the High Street, three other fires were burning in different shops at the same time, and for a while the Brigade was at full stretch.

The Bombing Intensifies

The bombing incidents increased in both size and severity of outcome during March and April. A more sophisticated device of about 15 lbs of gelignite was used on 26 April when it was electronically detonated on a cluster of ship-to-shore oil pipelines in the harbour estate. It took 30 firefighters with six fire engines to extinguish the resulting fierce fire. On 27 April, a considerable amount of damage was caused to the forensic laboratories in Verner Street when three bombs, one at the main door and two others at windows, detonated and started a fire. While the Brigade quickly dealt with the incident, considerable damage was caused to the laboratory and the forensic work carried out there was badly disrupted.

During May the explosions began to kill and badly injure people. On the afternoon of 1 May, a 12-year-old boy found a copper pipe lying in the road in Ballymurphy. Not knowing it was a homemade explosive device he banged it on the road. It detonated in his hands causing injuries to his hands and face. On 3 May, a bomb containing approximately ten lbs of gelignite exploded at the Regent Bar in Dock Street, the second attack on this particular bar within a year. The following night a gas pipe in the damaged bar was set on fire by a crowd of people, who then stoned the firefighters arriving to extinguish the blaze. The police and army had to secure the area before the fire could be extinguished.

On 24 May, 18 people were injured, at least one seriously, when a bomb consisting of 15 to 20 lbs of gelignite demolished the Mountainview Tavern

---

[36] McKittrick, Kelters, Feeney and Thornton (1999).

at the corner of the Shankill Road and Cambrai Street. Local people helped firefighters in their rescue efforts and the injured were taken to hospital in a fleet of ambulances.

An explosive device was thrown into the reception area of Springfield Road police station on 25 May. A car pulled up outside the station and a dark haired man in his mid-20s got out with a suitcase, which he threw through the front door. There were a number of civilians in the reception area including two children. Sergeant Willits, who was on the scene at the time, 'realised what was about to happen' when he 'saw the suitcase hit the floor. He thrust the two children down into a corner and stood above them, shielding them as the 30 lbs of explosive in the suitcase went off. He was killed instantly but the children he'd protected escaped with their lives'. [37] Seven police officers, two soldiers and 18 civilians were injured and firefighters met the grim sight of debris, broken glass and blood as they arrived on the scene. Most of the injured civilians were walking past the station when the attack was carried out.

The bombings continued throughout the year. In the early hours of 18 June, a series of explosions badly damaged an electrical sub station on the outskirts of the city. The first blast caused a fire, which was being fought by firefighters when two more explosions occurred in another part of the plant, about half an hour after the first explosion. It was thought that this incident might have been an early example of booby-trap devices, which were designed to kill or injure members of the emergency services as they dealt with incidents; thankfully there were no injuries.

July continued where June left off. On 10 July, in the early hours, a 50 lb bomb exploded in an underground sewer at Ardoyne. It was believed that this bomb had been intended to destroy a nearby army post. When firefighters attended the scene to deal with a fire caused by the explosion they came under attack from a hostile crowd.

* *

At 9.00am on 12 July, following two years training as a Junior Firefighter, I reported for duty on Red Watch, Central Fire Station, in Chichester Street, somewhat nervous but keen to get started. That night there were a number of explosions in the city. The largest was caused by a bomb of between ten and 20 lbs of gelignite at the British Home Stores in Castle Arcade. The shop was badly damaged but thankfully no one was hurt.

---

[37] Styles and Perrin (1975).

The Regency Hotel

One of the worst incidents in 1971 was not caused by the Troubles, but was an accidental fire in the four-storey Regency Hotel in Botanic Avenue. On 17 July, at about 4.00am, the night porter of the hotel noticed smoke coming from underneath the door of the second floor bedroom, occupied by local television presenter and ex-Ireland rugby international, Ernest Strathdee. One of the people at the hotel described the bed being 'in flames' when he 'opened the door' of the bedroom, finding the room 'full of smoke.'[38] Firefighters from Chichester Street were in attendance within four minutes of receiving the call and in all seven guests were rescued by the Brigade. After 40 minutes the fire was under control but three lives had been lost. Mr Strathdee's body was found in his room and a Canadian couple in their 50s, who had the room above his, also died. Among those rescued by the Brigade were two daughters of the Canadian couple. All of those who died were killed by carbon monoxide poisoning from inhaling smoke, and the inquest heard that Mr Strathdee had been drinking heavily before the fire.[39]

Charlie Hughes remembers the incident:

> Ernie Strathdee fell asleep with a cigarette. We got the call at 4.00am. There were two Canadian people, a mother and a father; they were in the room immediately above Strathdee's room. They died; their bodies were the first bodies I had ever seen. I knew there was a body, but I couldn't see the body because he was lying over at the window in the foetal position, burnt to a crisp and all the debris on top of him. It was only when I got outside I realised my stomach was churning, it was really churning and I was sick. That was my introduction to the fire service. It was started by a cigarette: you could see the whole thing; the half moon shape in his mattress; the half moon shape in his suitcase underneath his mattress; the circle of fire between the half moon where the suitcase was and the half moon continued round on the carpet on the floor. And, as I say, the man and the woman, I always remember the man had a pair of horn-rimmed glasses on him. And they were both lying at the other side of their bed in the room directly above him.[40]

---

[38] *Belfast Telegraph.*

[39] *Belfast Telegraph.*

[40] Charlie Hughes.

My First Bomb

Despite the on-going incidents I had had a relatively quiet first couple of weeks. My watch were not on duty for some of the calls and for some of the shifts I was assigned to the foam tender, a specialist vehicle that didn't go out too often. This changed in the early hours of Friday 30 July when I went to my first bomb explosion. At approximately 3.30am we got a call to an explosion in Victoria Street, just around the corner from the station. When we arrived we found that an estimated 18 lb. gelignite device had detonated, blowing down the front window and one wall of an all night cafe, the Nite Bite.

Eleven people were in the cafe at the time; a number of them had been blown to the ground and were covered in dust and rubble. The worst injuries were to a woman who had been sitting on a stool at a breakfast bar adjacent to the wall. She had one leg crossed over the other, and the blown-in, collapsing wall had trapped her under the rubble from about the waist down. The blast had wrapped her leg around her waist and we could see that it had been broken in a number of places. In these circumstances the rescue is a delicate operation and we needed to carefully remove the rubble from her so that we could get her into the ambulance, without causing any more damage. We were able to remove the rubble from her and get her onto a stretcher and off to the Royal Victoria Hospital. The next day she was said to be 'fairly comfortable'. The ten other people in the cafe were also taken to hospital suffering from shock and minor injuries. The explosive device had been put in the hallway of the building concerned, which also housed the RAF careers office, and we assumed that this had been the target for the bombers.

Internment

August 9 saw the introduction of internment without trial, and thousands of soldiers were involved in early morning raids detaining 342 people from republican areas.[41] Many of the wrong people were detained and 104 were released within 48 hours. The result was an immediate increase in violence with gunfights and massive disorder becoming commonplace, the death rate rose steeply.[42] Rioting, fire-raising and bombing swept through the city, and

---

[41] Fay, Morrisey and Smyth (1999).
[42] McKittrick, Kelters, Feeney and Thornton (1999).

by 10.00am on the morning of 9 August it was impossible to estimate the number of factories and shops damaged by fires and bombs. In the early afternoon three buildings close to the city centre were set on fire with petrol bombs, Thompson Reid's in Joy Street, McCutcheon's in Cromac Square and Government Offices in Albert Street. Even though the rioting continued and lorries were being hijacked for barricades at the time, firefighters managed to extinguish all three relatively quickly. The worst hit area for rioting was in the east of the city, which was a scene of utter destruction. More than a dozen corporation buses were hijacked and used as barricades, some of them set on fire, and a number of buildings were also burnt. At 8.00pm four factories in the Short Strand/Mountpottinger area were burning fiercely and equipment from inside them was being thrown into the street. Burnt out buses blocked the entrances to numerous streets as the trouble spread and by midday practically every street corner on the Falls Road had become a petrol bomb factory. The Brigade was stretched to the limit and struggling to cope with the sheer volume of calls. That night approximately 240 houses in Farrington Gardens, Cranbrook Gardens and Velsheda Park, owned by private landlords, were destroyed by fire, while the families living there moved out. It would seem that someone had cut the gas mains and set the gas alight to start the fires. As was often the case, there were different reasons given for the destruction, depending on whom you listened to. The Fire Brigade was prevented from putting out many of the fires by armed men and the on-going gun battles in the area.

I wasn't on duty that night and, with some friends, I stood on the Crumlin Road at the top of our street opposite Everton School and watched the houses burning. I was itching, like most firefighters I suppose, to be able to attend and fight the fires I could see but not get to. Homes, shops and factories were attacked and set on fire and, in the 24 hours up to 5.00am on 10 August, firefighters had dealt with 57 fires, many of them serious, despite being attacked by rioters and on some occasions having their hoses cut while trying to fight the fires. On 35 occasions firefighters were unable to attend incidents due to hostile crowds and gunfire. Despite being involved in what was described by a local paper as 'dodge the bullets operations', as they avoided bullets, bottles and bombs, no firefighters were injured.[43] By the end of that 48 hour period Belfast looked and felt like a war zone, and, according to one official estimate, at least 2000, possibly more than 2500,

---

[43] *Belfast Telegraph*.

families moved out of their homes in and around August 1971 – many with just the clothes they were wearing.[44]

Ken Harper recalls that the 'service learnt on its feet during that time, everything we learnt was never taught to us but [instead] learnt through hard experience. Recruits became veterans in two years, whereas it had taken 20 years to get to that point before the Troubles.'[45] Bob Pollock describes the service as a 'big learning curve', kept strong by 'ex-servicemen who had gone through the war' and who 'steadied everybody.'[46]

There followed several nights of Troubles-related incidents, scattered throughout the city, bombing, rioting, incendiaries and major fires. To take just a few examples; on 20 August a malicious fire broke out in a store owned by British Paints Ltd. in North Queen Street. In the early hours of Saturday 21 August a major fire, caused by a bomb and, it was thought, incendiary devices, destroyed a large four-storey block of offices, warehousing and shops at the junction of Wellington Place and Upper Queen Street. Even though the Brigade was in attendance within a few minutes, it took more than an hour to bring the fire under control. At the height of the incident, ammunition and guns were moved from Nelson's sports shop on the ground floor, as there was a fear that the ammunition would explode in the heat. On the Sunday of the weekend that followed, Belfast's first strip club, 77 Sunset Strip, was severely damaged by a 50 lb. bomb. Just to show that their sense of humour wasn't damaged, the owners placed an advert in the papers the following night saying that 77 Sunset Strip, 'the most explosive show in town', was now closed.

On 25 August, a 23-year-old man was killed when an IRA bomb detonated at electricity board offices on the Malone Road. Thirty-five people, mainly typists and secretaries were injured. One of the injured was pregnant and one was a 20-year-old woman who would later be killed by a bomb at the Abercorn restaurant. The explosion caused a stairway full of evacuating staff to collapse and several people, including young secretaries and typists, were buried as tons of rubble collapsed on them. Many staggered dazed and bruised from the building as others lay screaming, their clothes in shreds and blood pouring from wounds to the leg, head and body. Fellow workers

---

[44] McKittrick, Kelters, Feeney and Thornton (1999).
[45] Ken Harper.
[46] Bob Pollock.

who escaped the blast pulled away the debris in an attempt to help their colleagues until the Fire Brigade arrived.[47]

On 2 September, just after noon, four bombs detonated within a few minutes of each other in the city centre. Thirty-seven people were injured, some seriously. Three of the bombs exploded at Bedford House, behind the City Hall, minutes after an explosion at Glengall Street. The detonations occurred at one of the busiest times of the day and there was panic as shoppers and the injured fled from the various scenes. On the night of 13 September, nine people were injured during a night of bomb attacks throughout the city. Two explosions on the second floor of the Wolfhill Mill started a fire, which severely damaged the building and put 250 people out of work.

Roger Dawson recalls:

> [Wolfhill Mill] was a lost cause, I had been there many times, it must have been a fantastic place in Victorian times, it had its own reservoir. It was a brick building with wood throughout, massive timbers with a Belfast roof. We tried to put it out and you were standing within 30 foot of the wall. Because there was a water reservoir you had a limitless amount of water. But you had an inferno, in which no amount of water would put it out. It was an absolute lost cause.[48]

The Build Up to Christmas

There was no respite and on 4 October a soldier died and five others were injured when a large bomb exploded at an army post in a terraced house at Cupar Street. The men were buried under the rubble in the aftermath of the explosion and soldiers and civilians helped firefighters in their rescue efforts, digging in the debris with their bare hands. To make matters even more difficult there were water restrictions at the time, due to a lack of rainfall and an explosion at a water pumping station. Lorries carrying 200 gallon water tanks, which enabled firefighting operations to continue, were accompanying fire engines. I remember we were quite amused that two of the tanks were labelled 'Guinness', if somewhat disappointed that they only contained water. On 12 October, while firefighters were working their way through the front window of a paint store on fire in Hamilton Street

---

[47] McKittrick, Kelters, Feeney and Thornton (1999).
[48] Roger Dawson.

with a jet of water, there was an explosion caused by a paint tin detonating in the heat. Station Officer David McLeery was blown out and into the street when, after spotting what turned out to be a bomb at the front of the building, he ran into the incident to warn the three firefighters who were inside. Soldiers at the incident were fired on from the Markets area as the fire was being fought.

My friends and I went to the Celebrity Club a few times in our teenage years. Situated above the C&A clothes shop in Donegall Place, close to the City Hall, it was one of the few places considered relatively safe to meet members of the opposite sex in the city at that stage of the Troubles. On 24 October it was crowded when a man and a woman, both armed with heavy revolvers, entered the foyer and held a group of people getting out of a lift at gunpoint, while a second man planted an explosive device. The bombers ran from the building, shouting that anyone who came after them would be shot. Two policemen, who were in the club in plain-clothes, ran out and shot at the bombers killing one of them. The bomb exploded two hours later causing a fire, which was quickly extinguished by firefighters, who were on standby at the incident. The bomb disposal officer had a lucky escape that night. He had already examined the bomb, had gone to get some more equipment, and was walking back towards the device when it detonated.

On 3 November three men entered the ground floor bar of the Red Lion at about 4.30pm. While one of the men held the people in the bar at gunpoint, the other two planted a bomb. Some of the customers left through the front door, but those who tried to leave through the side door found it had been locked due to fears that a bomb could be left in the toilets. Three people were killed and 36 injured. Many of the injured were buried in the rubble and firefighters, soldiers and civilians worked hard to try and rescue them. At almost the same time as the first detonation a second device exploded in a drapery shop on the other side of the police station, which was next door to the bar.[49]

There were many other bombs and fires that month. At lunchtime on 17 November, two young men, armed with a pistol, glued a 20 lb bomb to the counter of a furniture shop in Waring Street. A member of the bomb disposal team had a lucky escape when he fixed a fuse to the lid of the plywood container containing the bomb in order to blow the lid off, and left the immediate vicinity just seconds before the device detonated and wrecked

---

[49] McKittrick, Kelters, Feeney and Thornton (1999).

the shop, smashing hundreds of windows in nearby shops and offices. Bob Pollock remembers the incident:

> There was a furniture place in Waring Street and it was gutted, a real inferno inside, but we put a ladder up the side and we got in through the window. There was Lloyd Brown, myself and Sammy Gamble and as we got into it we brought a jet up, it was just like a little office, and we opened the door of the office and it was just an inferno below us. One of us lay down on the jet and just kept spraying it inside and the other two would sit back and get some air, then somebody would take over and you would sit back and then the other one would take over. And as I got off the jet I came back and there was a desk with a big long drawer on the desk and as I leaned on it the desk drawer came open and it was filled with money. I mean notes that deep, and just as I looked at this Peter Neeson came through the window. 'Well done boys, they were telling us there was money, give us that' and he took the drawer down and gave it to them.[50]

A Close Call

On 26 November, firefighters had a close call when they were fighting a fire in EAB Discount Store in College Square North. They had entered the building to fight a fire, which had been started by an explosive device, when a second bomb exploded, causing the building to partially collapse and trap them in the rubble.

Ken McClune remembers the day:

> My very first day on duty I managed to get myself blown up at EAB. The first bomb went off and we went in to firefight, because in those days you didn't wait for clearance. We were no sooner in there than the second bomb went off. When we arrived the front of the building was blew out, with a fire going well inside. Bobby Pollock and I were on the pump escape and the two of us went in with the jet as the fire was spreading after the explosion, and then it was just sort of a deafness. I knew I was being blown backwards but I couldn't do anything about it, just like a leaf in the wind. I was blown out of the building and landed out several feet away. I went

50 Bob Pollock.

back in looking for Bob, and the jet was just lying there, hosing water, and there were screams from the guys trapped under the rubble that came down on them. I couldn't do anything for them, as I was obviously on me own, so I grabbed the jet to keep the fire back, because the fire was beginning to spread again towards them. There were several other guys who got trapped under the debris. Roy Brown and I think Ivor Maidley was buried under that, Red Richie was the driver; Victor McAllister was Temporary Station Officer at the time, he was the officer-in-charge. Then the next thing we were relieved from some guys from fire safety. Jimmy Kerr, he came in and he took the jet off me.[51]

Bob also recalls the incident:

We had arrived up and I was a BA man and Ken McClune, it was his first shout, and we ran out the jet because a fire had started in it and the bomb had gone off. There was nobody else there so we assumed that that was it. So we got in, and the hose burst and I went back out to get a new line of hose and as I came back into the building there was a flash about ten feet from me and it was a secondary device and it blew. They reckon it was a 40lb bomb. And all the stour and dust and whatever was left came down on top of us, and the rubble. Some guys got trapped between washing machines. I couldn't see anything, I couldn't hear anything, and I had the torch we carried and I shone the torch in my eyes to see that we were alive and that I could see, and then the noise sort of came back and it was a very hard ringing in my ears. They took us up to the hospital and checked us out. They brought us into the old Officers Mess and they gave us a brandy each and because it was only half four, and as we weren't due to go home till six they let us home early.[52]

There were more incidents in December. On the 2nd, 21 people were injured when two bombs exploded in the city centre within minutes of each other. The first was in the Copper Rooms restaurant where two armed men planted the device and told staff and customers they had minutes to get out. A man working at the front of the restaurant at the time said: 'the wall just came down around us and one man was covered in the rubble. We just did not know what had happened. Then the whole place seemed to go up

[51] Ken McClune.
[52] Bob Pollock.

in flames.'[53] The second explosion was at an office equipment showroom in York Street. Nine people, including a baby, who were in a passing bus were injured and taken to hospital.

McGurk's Bar

The McGurk's Bar atrocity caused the biggest loss of life during the Troubles in Northern Ireland until the Omagh bombing in 1998. The bomb detonated at 8.47pm on Saturday 4 December in the busy bar in North Queen Street, reducing the building to a pile of smouldering rubble. Firefighters working at the locus, helped by ambulance workers, the security forces and local people, were pulling away the debris with their bare hands. Fifteen people suffered terrible deaths that night, eight from crush asphyxia, three from burns, two from multiple injuries, one from blast and crushing and one from carbon monoxide poisoning. The McGurk family lived above the bar and the owner's wife, Philomena, and 14-year-old daughter, Maria, were among those killed. Amongst the others who died in the explosion were 13-year-old Jimmy Cromie, a friend of the bar owner's son, and Philip Garry, a 73-year-old school crossing patrolman.[54]

Divisional Officer Benny Marks, speaking in the *Belfast Telegraph* on the Monday night, praised the army and civilians for their heroic work during the rescue operation, which extended into the Sunday morning. 'I must also pay tribute to my own firemen,' he said. 'On Saturday night they had a bloody initiation to this type of work and they performed heroically. Many of them are only young men but no praise could be high enough for them.' Mr Marks also thanked the 'willing hands' of the North Queen Street area for the help they had given. He said they had greatly helped in the work of extricating bodies; 'when we got them organised at the fringe of the pub, with the firemen doing the actual work of extraction, they were a very great help'.[55] He described the work that had to be done as 'nauseating and sickening'. Altogether, there were 25 firefighters on the scene. They worked in the rubble from 9.00pm on the Saturday night until 6.00am on the Sunday morning, when the last body was found.

---

[53] *Belfast Telegraph*.

[54] Gillespie (2010).

[55] *Belfast Telegraph*.

Fire Officer Leslie Johnston, in a television interview for UTV, given in the '80s remembers that it 'was very obvious' that 'there was very, very little hope for anybody trapped in the building... The main job there was to get those out that were trapped', recalling he 'can still see some of those people actually burning'.[56] A 14-year-old boy, Seamus Kane, who was playing snooker upstairs at the time of the incident, later said: 'Everything went dark and I remember being under the rubble. My back was scorched by the fire and I have stitches in my leg'.[57] Among those pulled from the rubble was ten-year-old John McGurk, who went on to become a respected journalist in Belfast. In an interview for *The Irish News* in 1996 he described events on the night of the explosion. He had been playing a board game with his older brother and two school friends between 8.00pm and 8.30pm and recalls 'it was like something out of a really bad horror film. I remember tumbling through air and space amid this massive rush of wind and noise. I couldn't remember anything else because I must have been unconscious for a while but I don't know how long it was. It was miraculous for me, a person, Jimmy, who was just a few feet away from me was killed. I woke up and I really didn't know what had happened. There was then the realisation that the building had collapsed and I was stuck. I wasn't physically injured. The only injury I had was an injured finger'.[58]

The Salvation Army

Most firefighters from Belfast had a soft spot for the Salvation Army, not necessarily for any religious reasons, more because when we were out at incidents in the city during the difficult days and nights of the Troubles, they would turn up in a car or a van with tea and filled rolls:

> I can remember coming on duty at night, on night duty at 6.00pm and not getting back in the station until half eight the next morning. If it hadn't have been for the Sally Army we would have starved to death. They used to come out to bombs and feed us. Hot tea and coffee and a bit of a bun or a bit of cake.[59]

---

[56] Forgotten Heroes.

[57] Fay, Morrisey and Smyth (1999)

[58] *The Irish News*.

[59] William Hoey.

When I see the Salvation Army in the street, I always dig deep in my pocket because they really fed us. We were leaving the station at maybe half six at night and never got back till after nine in the morning. We never got back to the station at all, you were just going from one to the other. And they fed us, the Sally Ann, it was always great to see their wee wagon coming with the tea and whatever else.[60]

For this reason, most of us have difficult feelings over what happened on 6 December. At 2.00pm two gunmen, one armed with a submachine gun and the other with a pistol, planted three explosive devices on the ground floor of Munton Brothers shirt factory on the Dublin Road. The subsequent explosions caused a massive fire in the traditionally built, mill style brick and timber factory, described by Benny Marks at the time as 'one of the biggest since the Blitz'. At its height more than 70 firefighters, including crews from as far away as Donaghadee, Glengormley and Larne fought the fire. Ten fire engines supplied 12 firefighting jets and four ground monitors. A turntable ladder and a hydraulic platform were also in use. A desperate attempt was made to prevent the fire spreading, by creating a 'wall of water' between the building on fire and nearby premises across Marcus Ward Street. However, despite this massive effort, the fire spread to the other side of the street, as the water being used was turned into steam by the heat of the fire. At one stage the firefighters were forced to withdraw leaving only ground monitors in use. Eventually the Brigade managed to bring the fire under control after more than three hours of punishing and dangerous work. The next day, small fires were still smouldering in the destroyed building. Hundreds of people lost their jobs and the fire was said to have cost a million pounds.

The Salvation Army had a citadel adjoining the factory building and, as was their usual practice throughout the Troubles, they supplied welcome cups of tea to the fire crews dealing with the incident. Among those helping to do this was Mary (Mamie) Thompson and her husband Sam. As the fire spread they were also involved in salvaging furniture, instruments and fittings from their citadel. A wall of the citadel collapsed on them as they carried out the task, killing Mrs Thompson, described by her husband as, 'a happy wee woman, one of the best', and injuring her husband and four other men. Following this horrendous incident, conflicting accounts emerged as to whether they had been warned to stay out of the building following the

---

[60] Dessie McCullough.

explosion. Salvation Army members said that firefighters had told them it was safe to go in, but senior police and fire officers said they were unaware anyone was inside the building. A senior policeman said that firemen had given permission for the Salvation Army to salvage property from a minor hall adjoining the citadel, but that no one had asked his permission to go into the citadel itself. Benny Marks, speaking the day after the fire said: 'We did not know there was anyone in the building. If we had been asked we would not have given permission – it was a most unfortunate accident'. During the inquest the Coroner told the jury that Mrs Thompson had gone into the building of her own volition, and the inquest returned a verdict of misadventure. In January 1972 Eileen O'Brien interviewed Mr Thompson for *The Irish Times*. He said that he had played in the band since he was eight years old and had gone to salvage instruments. His wife had insisted on going with him and she had made tea for the firefighters. 'I tried to clear up what I could. Then the whole building fell right around us. We had no family so we always went everywhere together. Now I am by myself.'[61]

This is an incident that remains with the firefighters who were there:

> I was driving the Chief that day. Full attendance, the next thing was a make-up, make pumps ten or something I think it was, and the Chief he came out and he wore a big black oilskin coat, the coat they had for salvage. I went to give the boys a hand. It was jumping Bankmore Street, heading towards the BBC. When we arrived up it was coming through on three floors, the flames were coming out of every window.[62]

> Bobby Malcolm and I were setting up a ground monitor in the street opposite, it was going from top to bottom, and that fire jumped the street. We were setting up a ground monitor and we were watching the front of the building because the heat was intense and there were cracks and there was this noise. You've never seen three firemen take off as quick up the street towards Great Victoria Street, Bobby Malcolm leading the charge. But we went back and fought the fire. Now the Sally Ann were trying to rescue what they could, they had got their piano and all out down the alleyway. The Sally Ann had sent their van round. I had just got

---

[61] McKittrick, Kelters, Feeney and Thornton (1999).

[62] Jimmy Armstrong.

a cup of tea in my hand and a wee bun and the next thing Jim Graham, he came running down and says, 'Blanket, blanket quick', and I turned round and said, 'What's wrong'. He said, 'There's a collapse up here, there's a woman caught'. So I set the tea down and ran back up with him. We went down to the hostel and the whole hostel had collapsed and this wee woman was just inside the door and she was buried, the whole wall along that side had fallen. That wee woman was killed.[63]

There were at least 30 more bombs in the city that December, several causing fatalities, injury and/or major fires. On Saturday 11 December, just one week after the McGurk's bar bomb, there was a 'no warning' device planted at a furniture showroom on the Shankill Road. Two men and two small children were killed and 19 others were injured, some of them seriously. The two children who had died were being pushed past the showroom in a pram when the explosion occurred and they were buried in tons of rubble. Firefighters from Ardoyne station were quickly on the scene and they directed local people in the rescue operation, digging with their bare hands in the smoking debris. Many of the injured lay trapped and a digger was used to clear some of the rubble. Army lorries were brought in to help clear the wreckage and emergency teams treated the injured.

Another one that I remember specifically was a furniture shop on the Shankill Road. The building had come down on, I didn't know how many people, but at the end of the day I was digging away and I pulled a little pushchair out and there was a young male child in a little blue coat. He was approximately the same age, a few months old, as my son who also had a little blue coat exactly the same. I found that quite harrowing at the time.[64]

The child in the blue coat was 17-month-old Colin Nicholl. His adoptive mother Anne, who was not with him at the time, had watched a news report of the incident. She was interviewed for television after the event and said,

I saw it on the television and I said to the sister, 'That's terrible,' you know babies getting killed. I didn't realise it was Colin and

---

[63] Charlie Hughes.
[64] Jimmy Hughes

I'll never to the day I die forget seeing that fireman with the baby wrapped in a blanket. You know it'll always be in my memory. Jackie went on to say; 'it's hard to explain in words just how much we do miss him, you know. The great tragedy of it all is that both Anne and I, we just can't understand that he's away from us and we won't see him again. Through all the tragedies you don't realise how much it's going to hurt.'[65]

---

[65] McKittrick, Kelters, Feeney and Thornton (1999).

PART TWO

1972 to 1976

The Blood and Screams of Daily Life
in Belfast[66]

1972

THERE SEEMED TO be a terrible inevitability to the realities of life for the Brigade at the beginning of 1972. We knew that there was no end to the Troubles in sight, but we didn't know just how bad things would become. The year was one of turmoil, violence, destruction and political confusion. The terrible toll of death and injury was much worse than in previous years and we had to deal with atrocities that would have an effect on us for the rest of our lives. January of the new year began very much as December of the old year had ended and the first day of the new year saw seven explosions in the city.

On 2 January, I was on duty in Chichester Street and at the start of our day, following parade at 9.00am, we checked our appliances and gear and carried on with the day's duties. At about 1.00pm we sat down to lunch in the mess when we heard the familiar loud bang of a nearby explosion. A device had detonated in a beer lorry in Callender Street, just a few hundred yards away. The explosion, intensified by the narrowness of the street and the height of the surrounding buildings, blew the lorry's load of bottles in all directions and over 40 people were injured, mainly by flying glass. Some of the injured were in nearby shops when the windows were blown in on them. We were on the scene very quickly, due to the close proximity of the station, and when we arrived we were met with dozens of injured and shocked people. There were trails of blood on the street where injured and bleeding people had run from the area. A second lorry, behind the one that

---

[66] Keane (2011).

exploded, was carrying a load of Harvey's Bristol Cream Sherry and it was set on fire by the bomb.

Bloody Sunday

'Bloody Sunday' in Derry/Londonderry occurred on 30 January when 14 people were shot and killed by paratroopers during a protest march against internment. This was considered by many as one of the most significant events of the Troubles, and it can be seen as a watershed event for Northern Ireland as a whole. It had enormous ramifications, not only because 14 lives were lost, but because attitudes were hardened significantly, paramilitary recruitment increased and much more violence and many more deaths were generated.[67]

The increase in violence was immediately felt in Belfast when, on 31 January, an estimated 50 to 100 lb. bomb was left in a cleaner's van at Castle Arcade. The bomb caused a small fire, which was quickly dealt with by the Brigade who also helped with the injured. There were two other bomb explosions at about the same time. One of which exploded in Great Victoria Street after a man had lifted a device from his office and carried it out into a passageway before it detonated. Overnight violence erupted in the city. A soldier was shot and seriously wounded in the Falls Road area. Nail bombs were thrown and there were gun battles in a number of places. More than 30 vehicles were hijacked and set on fire and several public houses and offices were bombed or set alight. The Broadway cinema also suffered a fire and was seriously damaged, as were a number of other business premises.

In February the bombing, rioting, shooting, nail bombing, hijacking and burning of vehicles continued throughout the month, and malicious fires destroyed a number of buildings. On 23 February, a 500 gallon oil tanker was hijacked by two armed men at about 7.30am. A short time later a bomb exploded in the cab of the tanker but the oil didn't ignite. Roughly 15 minutes later a second charge detonated under the tanker. On the morning of 25 February, three devices detonated in the city within an hour of each other. One of the targets was a music shop in Wellington Place; two men planted a bomb and, having sprinkled petrol over the premises, gave staff five minutes to get out of the building. A fire, which quickly took hold of the four-storey building, followed the explosion. Firefighters prevented it from

---

[67] McKittrick, Kelters, Feenet and Thornton (1999).

spreading to adjoining premises by applying cooling jets of water prior to extinguishing the fire.

The Abercorn Restaurant

On Saturday 4 March 1972, I was in the city centre with my then girlfriend. As I remember it, it was a cold but bright afternoon. At about 3.30pm we decided to go for a coffee and thought that the Abercorn might be a good place to have a seat. However, the city was busy, particularly in the Cornmarket area, and when we got to the Abercorn it was pretty crowded so we decided to look elsewhere. I have often felt lucky that we did so, because about an hour later, two young women left a bomb in a bag under a table in the Abercorn restaurant. At 4.30pm, two minutes before the device exploded, police received an anonymous 999 call to the effect that a bomb would go off in five minutes. Just after 4.30pm the device detonated killing two young women and injuring over 70 people, many of the injured were badly mutilated.[68] The bomb had been left at the back of the restaurant and when it detonated the blast swept through the room, closely followed by a fireball.[69]

The two young women, Ann Owens and Janet Bereen, who were killed had been waiting to be served when the explosion occurred. Ann Owens worked for the Electricity Board of Northern Ireland and had been injured at an earlier explosion in the board's headquarters. Janet Bereen's father was a senior anaesthetist at the Royal Victoria Hospital. On the day of the explosion he took part in operations on the victims, unaware that his daughter was among those killed, her body was brought in as he worked to help others. Among the injured were Rosaleen McNern and her sister Jennifer. Rosaleen was a 22-year-old secretary and Jennifer was a school meals supervisor. Rosaleen was to be married that year and the sisters were in town to shop for clothes for the wedding. Sitting near the girls was Jimmy Stewart, aged 36. Rosaleen lost two legs and an arm and Jennifer lost both of her legs. Jimmy Stewart also lost both legs. Firefighters from Chichester Street station attended the incident:

> The other one, which really sticks out, is the Abercorn. It was a
> Saturday afternoon. We were sitting having our tea when the call

---

[68] McKittrick, Kelters, Feenet and Thornton (1999).

[69] Hennessey (2007).

came. We went round and there was a poor guy up against the coffee bar, on the floor, minus legs from the knees down. He was tying his own legs together with a tea towel.[70]

When we arrived there was no fire, just a lot of smoke and in quite a narrow street. The door and all the windows were blown off, we could hear crying inside. As we went in we could see people lying about. There were a lot of people very badly injured. I mean, a guy had lost both feet, there was two women, two sisters, one lost an arm and two legs and the other one lost two legs. There was a girl got her eyeball split, right through. Another girl, there's photographs of Lloyd Brown carrying her with a table leg through her thigh. But we got in and we started carrying them out and we were walking over this door, which was rocking back and forward. We carried out the guy, the guy who had lost both feet was conscious, we carried him out and there was no room for him in the ambulance, we had to put him on the floor of the ambulance, and then I went in and I picked up his feet and there was, I don't know if it was a towel or a curtain or something, but I wrapped them in it and set them beside him in the ambulance. That was very traumatic. When we eventually got in and were looking around somebody happened to lift the door and there was a body under the door, a girl under the door had been blown to pieces.[71]

On 2 August 1972, Rosaleen McNern was married to Brendan Murrin in the Donegal village of Killybegs. Jennifer McNern was one of the bridesmaids. Jennifer and Rosaleen were taken up the aisle in wheelchairs. On 5 October, Jimmy Stewart married Florrie Orr at the Welcome Evangelical Church in Cambrai Street, Belfast.

Rosaleen said:

You don't get past your luck. There is an old saying, 'If you are going to drown you will never be hung'. If it's for you, it's for you. If we were not sitting in that restaurant it would have happened to someone else. I don't feel cynical, and I don't feel bitter about the people who caused the explosion. It's more of a curiosity. I would like to know why they did it. I sometimes wonder if they

[70] Ken McClune.
[71] Bob Pollock.

thought they were right. Somehow I think the people who did it will feel worse than us in the end.[72]

Despite the widespread revulsion that followed the Abercorn bomb, the shooting and bombing continued in the city. At about 3am on 7 March, four men, at least two of them armed, entered the biggest department store in Belfast, the Co-op in York Street, and held the night security guards at gunpoint. The men stayed in the building for some time and, when they released the security guards, they told them to leave the building. At about 3.50am there was an explosion on the second floor that started a fire, which quickly spread to the third floor. Firefighters attacked the fire as the sprinkler system actuated and the fire was quickly contained. A 40 lb. bomb was found later that day in the store. Bob Pollock recalls that 'the first time we had gone into the Co-op they had put in a bomb and they had also put a bomb with a trip wire, and we all went in and every one of us stepped over that trip wire, everyone'.[73]

On 14 March, a 200 lb. bomb, thought to be the biggest bomb in Belfast since the Blitz, was planted in a Coca-Cola lorry at Sandy Row. Firefighters were still at the scene the next day, clearing wreckage from severely damaged shops, covering roofs with tarpaulins and ensuring the buildings were safe. The damage extended over a 200-yard radius.

Lower Donegall Street: A Car Bomb

On 20 March, I was on duty at the fire station in Chichester Street, it had been a relatively quiet morning for us but just before midday we heard the loud rumbling bang of a big explosion, which shook the city. Almost immediately we were turned out to Donegall Street. When we got there a scene of absolute devastation confronted us, with casualties scattered around in the dirt and dust that is inevitably caused by a blast of this sort. As a result of the explosion a huge ball of fire had roared down the street and the street looked like a 'battlefield'.[74] Victims were everywhere, they had blood pouring from wounds, some had their clothes blown off by the force of the explosion and many were just staggering around. Panes of shattered glass had blown onto the street injuring dozens of bystanders and hundreds of people were in shock. A teenage

---

[72] *Belfast Telegraph.*

[73] Bob Pollock.

[74] *Belfast Telegraph.*

girl had her legs blown off and it was obvious that there were a number of people dead in the street. Towels from a beauty salon were used to dress the wounds of some of the injured, and blankets and tarpaulins were used to cover the bodies of those obviously beyond any help. While shouting, screaming and confusion was all around us, our training and experience seemed to kick in, even for relatively young firefighters like myself, and we soon started to carry out our duty, helping the living, making the locus and its surroundings safe and dealing with the dead. I assisted ambulance staff with the injured and was one of those who carried away the mutilated body on one of our tarpaulins, while other firefighters did much the same.

The bomb, estimated as approximately 100 lbs, exploded in a Cortina car, following contradictory telephone warnings about a device in the area. In a horrible twist, the driver of a bin lorry nearly prevented the tragedy. At the inquest he said that the car had pulled up in front of his vehicle in Lower Donegall Street and the driver, conspicuously avoiding eye contact with him, rushed away from the scene. He warned a number of people, including his supervisor, that there may be a bomb in the car but at that moment a coal lorry hit his lorry, spilling coal onto the road. The lorry driver started to help pick up coal from the street and his attention was diverted from the car. A short time later he was continuing to empty bins and he parked his lorry close to the Cortina. He wept as he told the inquest what happened next:

> My three workmates went to the back of the vehicle to empty
> bins, when suddenly there was a loud explosion. The next thing,
> I was lying on the roadway. I got up and went to the back of the
> lorry, but I couldn't find the foreman. One of my workmates was
> lying on the ground shouting, 'Help me, help me!'[75]

He suffered eye, back and nerve injuries. Seven men lost their lives in this incident, and 150 people were injured. Three Belfast Corporation workers were killed while working at the back of their bin lorry, another man was killed while driving along the street beside the lorry and two police officers were killed as they were examining the car. The final victim, 79-year-old Henry Millar, died in hospital two weeks later from injuries he received in the explosion.[76]

---

[75] McKittrick, Kelters, Feeney and Thornton (1999).

[76] Gillespie (2010).

On 29 March, Chichester Street Fire Station was evacuated due to a 150 lb. bomb in a lorry at the law courts next door. The ATO started to defuse the bomb by burning off the explosives, however, the device detonated and firefighters were required to deal with two cars on fire, together with a number of fires in both the Magistrates Court, adjacent to the fire station, and an indoor market a few hundred yards away.

In April there was a worrying new development when, a new type of chemical explosive was used to create a major explosion in Church Lane on 1 April. This explosive was made of fertiliser and was more easily obtained compared to the gelignite which had previously been in bombs. The bomb weighed about 150 lbs and was hidden in a coffin on a hearse and, when it exploded, virtually every window in the street was blown out and all of the buildings near the hearse were badly damaged. The heavy rate of shooting, rioting and bombing continued and, on the night of 18 April, Station Officer Alan Wright was shot in the hand as he was fighting a fire in a derelict building on the Grosvenor Road. The weapon used was thought to be a Thompson submachine gun:

> I immediately felt a thud on my gloved right hand, which was at that time holding the string of the ladder six inches away from my head. I certainly heard and felt the whoosh of bullets passing over and around me. As I threw myself headlong down onto the pavement, I remember shouting, 'I'm hit'. If I could have crawled into the joints in the paving stones at that moment I would have. As I lay there hugging the pavement I felt the rest of the volley whoosh over my head. The only thing I could think of was, is the next one going to go up my backside. I felt completely naked and vulnerable. I was frozen to the ground and afraid to move. At that moment big Charlie jumped down from the ladder onto the ground and grabbed me by the scruff of the neck and with me half crawling and he half dragging me we made it into the shelter of the derelict shop that we had just extinguished the fire in.[77]

On 25 April, a fire caused by a parcel bomb badly damaged a large warehouse near the peace line at Northumberland Street. The parcel was delivered by lorry and signed for, but an invoice clerk at the warehouse

---

[77] Wright (1999).

thought there was something suspicious about the parcel and raised the alarm, undoubtedly preventing casualties. Flames quickly took hold of the two-storey building following the explosion and the chemicals stored there caught fire. Firefighters fought for over two hours to bring the incident under control. Dense black smoke could be seen from miles away and exploding cans, and the chemicals they contained, hampered firefighting operations. Firefighters took a lot of smoke and could be seen gasping for breath with their faces blackened, although none of them required hospital treatment following the incident. In all, six fire appliances were used to fight the fire.

The Co-op Fire

One of the more spectacular incidents, and certainly the biggest fire I had attended up to that time, occurred on 10 May when the IRA attacked one of the most prestigious targets in the city, the recently constructed Co-op in York Street. I was on duty in Chichester Street and was on the emergency tender, which meant that although the 'Co' was in Whitla Street's area I was on one of the first appliances sent to the incident in a so-called 'full attendance'.

When we got there we could see that a device had already detonated and smoke was billowing from the third floor of the massive building. I was still 18 and, having been operational for less than a year by this time, I was relatively inexperienced by Belfast standards. I have to admit I felt a certain amount of excitement as I ran down the street to get my instructions from the officer-in-charge. In the rush to get into action I jumped out of the machine without my helmet, but luckily Billy Little, a wiser, calmer, older hand was on the same appliance and he shouted over: 'Get your helmet on!' Having quickly rectified that somewhat embarrassing situation, I was instructed, with another firefighter (I am pretty sure it was Billy Fraser) to take a jet up to the third floor and try to prevent the fire from spreading through the building. The practice at the time was to get the jet close to where you intended to fight the fire before the water was turned on, so you didn't have to drag the heavy water filled hose too far, an almost impossible task in some situations, and that is what we did.

So we were there, just below the third floor, when one of the oil tanks above us ruptured, and burning oil started to run, to my surprise fairly slowly, down the staircase towards us. When it got to the landing above us, and the wooden handrail also started to burn, our fear of telling the Station Officer we had failed to do as he instructed was overcome by our fear of the fire. 'Do you think we should leave?' Billy asked. 'I will if you will', I replied, and at that we ran down the stairs and out into the street, where we were met by our relieved Station Officer, Tommy Everett. What we didn't know was that there had been another explosion, thought to be a booby-trap, which had split the oil tank, and our colleagues had been sounding the evacuation signal (repeated loud blasts on an Acme Thunderer whistle). Tommy had thought he had lost us. In those days your station officer was a major figure of authority and hardly spoke to you unless you were in trouble. I don't know if Tommy was embarrassed by the relieved hug he gave us but we certainly were.

Two of our colleagues, Divisional Office Leslie Johnston (Jonty) and Sub Officer Jack Warden (Big Jack) were trapped inside the building by the fire. Following reports of people still being in the building, they had decided to go in to try to rescue them. Unfortunately they got cut off by the fire and had to follow the only option left open to them, making their way up to the roof, where they called for help. Even though no one could hear them over the noise of the fire, Alan Brown, the driver of the turntable ladder, who had the ladder up with Ken Spence at its head playing a jet of water onto the fire, saw them:

> I noticed someone on top of the building, I took another look and it was two of our hands, two of our men, five storeys up, frantically waving. You could only see them as the wind blew the smoke and the flames away from them. So I realised I had to stop what I was doing as a water tower and try to rescue them.[78]

Alan had the turntable ladder connected to a hydrant in order to direct a jet of water onto the fire and he had to quickly disconnect from the water supply, bring the ladder down and move it to a position where he had a chance of getting to Jonty and Jack. As Jonty put it:

> So there I'm sitting on the parapet of this building wanting to get off, smoke swirling around and the head of the turntable ladder about six to seven feet away from me. And I have to give credit where credit is due: Alan Brown whipped that ladder. In other

---

[78] Alan Brown quoted in Forgotten Heroes.

words he drew it back nearly to the vertical and slashed it in towards the building, in doing so it came to within two to three feet of me. I jumped, and caught the underside of it. I knew I was OK then, and then I could start thinking about my colleague, who I hadn't given a lot of thought to then. Self-preservation is a hell of a thing, so it is. And when I got myself secured I shouted Warden's name. Stick your hand out Jack! And out of the smoke came the hand. Well with my weight on the head of the ladder it kept it into the building so I could grab the hand, that was the whole story.[79]

At the height of the fire there were 70 firefighters with 11 fire engines trying to extinguish it, but in the circumstances we couldn't bring it under control. Soon the whole building was alight and we were reduced to preventing the fire from spreading through the covered bridge across York Lane and into the adjoining block. The column of smoke could be seen everywhere in the city as it rose hundreds of feet into the air, and we lost one of Belfast's most iconic buildings. After the fire the building was found to be unsafe and it was demolished. The cost of the damage was estimated at about £10million and over 750 people lost their jobs.[80] *The Irish News* offices were next door to the Co-op and the next day it failed to publish for the first time in its 160 year history, due to the damage caused by the fire in the adjacent building.

Jack Warden would eventually work with me in Ardoyne, but in those days he lived just above me off the Crumlin Road and, because he had a car, he would give me a lift home. That night as we travelled home he told me that he had just about given in to the smoke and had fallen to one knee to die, when he saw the head of the turntable ladder through the smoke: 'Brian, I never saw a better sight in all my life than the head of that TL.'

The bombing, burning and killing continued and at one incident, on 13 May, an explosion was prevented from being an even bigger problem by firefighters when they attended a bomb blast in a paint store in the Markets. Two armed men, who gave a ten-minute warning, planted the device. When the bomb detonated a few minutes later it blew out both the front and back walls of the building, throwing dozens of paint tins into the air and starting a fire. However, the bigger risk came from the 20,000 lbs of highly

---

[79] Leslie Johnston quoted in Forgotten Heroes.
[80] Parkinson (2010).

flammable and, in the right circumstances, explosive, cellulose paint being stored in the rear of the premises. A pall of thick, black smoke rose over the city as firefighters got to work with jets of water. Despite the risk of further explosions they fought their way through the smoke and burning paint, which at one stage ran burning into the street, to extinguish the fire and prevent it reaching the cellulose.

JP Corry's: Firefighters as Hostages

JP Corry's was, and is, a timber yard based on the Springfield Road. Due to its location – adjoining a republican area – and the fact that there was a military observation post on the premises, the yard was often a target during the Troubles. One of the most difficult fires for the Brigade to deal with occurred in the yard at about midnight on 18 July. In an attempt to force troops out of their observation post, gunmen and petrol bombers attacked the yard, starting a major fire and causing an estimated quarter of a million pounds worth of damage. The fire spread to two adjoining houses, making their occupants homeless. At one stage two appliances and their crews, who were trying to deal with the houses that were burning, were taken hostage by gunmen who demanded that the Brigade withdraw and that the families who had been made homeless by the fire be re-housed. To reinforce their demands the gunmen threatened to blow up the appliances and their crews unless the Brigade withdrew from the yard. They also said that any firefighters continuing to fight the fire would become targets for the gunmen. The Brigade withdrew from the incident and the crews were eventually released. The attacks on the yard continued and the intensity of the fire increased. The Brigade returned to fight the fire once the army had secured the area, but shooting continued throughout the firefighting operation and it was well into the next day before the fire was extinguished. Dicky Sefton was one of the hostages:

> Two fire appliances started to put out the bungalows. The next thing the IRA contacted Control and told them that if they didn't cease firefighting operations in Corry's Timber yard the firemen would be shot. So they ceased all operations and withdrew and the timber yard flared up again. I was a hostage until 4.00am and then they released the two fire crews and then released me half

an hour later and told me there was a bomb on board one of the fire appliances.[81]

Bloody Friday

Friday 21 July 1972 was a fine summer's day. I was on duty at Chichester Street fire station and had been rostered for the pump escape along with Ken Spence. This fire engine had a removable 50-foot ladder fitted with wheels to enable quick deployment for rescues from upper floors. It also carried the station officer so it always went first to any incident, other than known small fires. For this reason I was expecting a relatively busy day. After a reasonably quiet morning we got a call for a full attendance to the Albert Bridge, where we found an explosive device in the supports under the bridge. It was while we were 'standing by' at this location that the carnage started, and we heard the unmistakable rumbling bang of a bomb, quickly followed by many others, so many we lost count. Soon there were several palls of smoke over the city and we started to get our appliances drawn away to deal with what was going on. The pump escape was the last one to be taken from the stand by and committed to the mayhem. Ken remembers:

> I was 200 yards away from the main event... really we were screened from that (the Oxford Street Bomb) but the rest of the guys, who were in the station, weren't screened and they took the full force of what was going on, in terms of the bomb and in terms of the trauma. What we had seen that day was enough for anybody you know.[82]

In all, 20 devices were detonated in just over an hour. Nine people were killed and 130 were injured. A car bomb at Oxford Street bus station killed four bus workers and two soldiers, and two women and a schoolboy were killed by another car bomb on the Cavehill Road. Some of the bodies were so badly mutilated in these two explosions that it was initially thought a greater number of people had died.[83] The targets included Smithfield Bus Station, railway terminals in Great Victoria Street and York Street, Queen's Bridge, an M2 flyover motorway bridge, railway bridges at Botanic Avenue,

---

[81] Richard Sefton quoted in Forgotten Heroes.

[82] Ken Spence.

[83] Gillespie (2010).

Windsor Park football ground, Finaghy Road North, gas department offices, garages in Donegall Street and the Upper Lisburn Road, the Brookvale Hotel, a bar in the docks area, a taxi office on the Crumlin Road, an electricity transformer in Salisbury Avenue and shops on the Limestone Road. It was 'impossible for anyone to feel perfectly safe... there were cries of terror from people who thought they had found sanctuary but were in fact just as exposed as before'.[84] A Fire Brigade spokesman said: 'everyone was stunned and shocked to see such carnage, but our men did a great job. The control room was chaotic, and in the streets our appliances were hampered by traffic jams, which got worse by the minute'.[85] A woman who had just crossed the Cavehill Road when a bomb detonated said, 'there was flame and then when they seemed to have died away, there was nothing but only glass and blood'.[86]

Firefighters, as well as assisting the wounded at several bombs, also had a large fire to deal with as one of the explosions set fire to premises in Donegall Street. Hundreds of people were evacuated as firefighters fought the fire, attempting to stop it spreading to the nearby *Irish News* offices. A second difficult fire was caused by a car bomb at a petrol station on the Lisburn Road. Firefighters fought the fire but were driven back on several occasions by the intense heat and small explosions, caused by the petrol, before they were finally able to extinguish it. The scene is described by Parkinson in a section of his book entitled, *When Belfast Shovelled Up Its Dead*:[87]

> The sheer intensity and scale of this 'IRA blitz' was unprecedented, and the task of the authorities was not aided by the high number of hoax bomb scares. In the city centre the security forces started sealing off streets identified by callers, but as the calls increased almost by the minute the opportunities for fleeing office workers and shoppers to escape were reduced, as numerous streets had to be cordoned off. Hysterical and confused secretaries and

---

[84] Parkinson (2010).

[85] McKittrick, Kelters, Feeney and Thornton (1999).

[86] *Belfast Telegraph*.

[87] a title written in reference to photographs and a newsreel showing a firefighter, John Hyland (known as Captain Friendly due to his friendly demeanour and general decency), dealing with an armless, legless torso that had been mangled by the explosion in the Oxford Street bomb, using a shovel to move the remains into a body bag.

shoppers ran up and down streets showered with glass from broken windows, and fires broke out in many premises following explosions. Ambulance and fire crews, negotiating their way with difficulty through traffic jams soon made their way to the dead and injured, as thousands tried to make a speedy exit from the city centre. [88]

Bus and train services were cancelled and people could be seen lining the pavements, frantically hitching lifts out of the city. Soon a thick pall of black smoke hung over Belfast, along with the regular thud of a bomb exploding. Although police, ambulance and fire crews were hampered in their efforts by crowds of people running through debris littered streets and escaping motorists blocking main roads, they emerged with considerable credit for their efforts that Friday afternoon. Several reports referred to the professional approach of the emergency services. Smith's report in the *Observer* also entitled, 'When Belfast shovelled up its dead' focused on the casual 'matter of fact air' of the fire service workers who 'picked up the dead in big wide shovels until they got to the smaller pieces when they used their hands'. The *News Letter* reporter observed: 'the tell-tale emotional giveaways that manifested themselves in the drooped heads, the sloped shoulders, the drawn faces and the occasional tear'. The bus station manager was leaning against the 100 lb. car bomb when it detonated. He later told the inquest that he had just walked up to the car along with two soldiers and put his hand on the bonnet: 'as I leaned across the car it exploded. The blast hurled me through the air and blew off my clothes. I finished up on the roof of the offices. There was smoke and dust everywhere but someone pulled me clear and saved my life. I was severely burned and had bad injuries to my legs and body'.[89]

Firefighters from Chichester Street were there to help:

> The bomb went off in Oxford Street. It blew us off our feet in the yard. We ran out anyway, we had hose out of the hose store, started lifting up bodies and stuff.[90]

> The most serious incident I was at was probably the Oxford Street bombing on Bloody Friday, and I wasn't even in firefighting

---

[88] Parkinson (2010).

[89] *Belfast Telegraph*.

[90] Jimmy Armstrong.

kit. I had just been made a Leading Fireman in Fire Safety and I was on my way back from teaching people fire drills in Donegall Street when, as I turned round to face Musgrave Street, the whole of the bus station just blew out in front of me. I had heard another bomb going off somewhere else behind me and at that time we were leaving our firefighting kit under our desks to man the spare pump, so I was making my way back as quick as I could to get changed into my kit. As I went I couldn't ignore a man who was lying in the middle of Oxford Street and he had his arm up and he was calling and I went over and I took him in my arms and held him until the Fire Brigade and the ambulance arrived. I was trying to comfort him but I could feel rattles coming from his body. He obviously had bones broken but I helped him up onto the stretcher and put him into the ambulance along with the ambulance crew.[91]

A group of junior firefighters had just finished two years of training and were expecting to be taking part in their Passing Out Parade that day:

That was my last day in the school. We were polishing all our gear; the appliances were shining, ready for us to have a passing out parade or inspection. We knew that there was something going on in Belfast because we could hear explosions going off. Central's appliances were out and those were the appliances we were going to use for our passing out parade. An explosion went off pretty close to where we were and the Chief Officer at the time, Robert Mitchell, he came in and said 'look guys your day has been cancelled, I would advise you to come away from the windows because there's a lot of stuff kicking off out there'. So we had just walked out of the classroom when the Oxford Street bomb went off. It was just across the road from where we had been standing. As we walked out of the classroom door the glass came flying in around us, the slates came rolling, sliding off the roof. We ran up the outside balcony, down the stairs and stood in the yard, a bit shaken up. So we were really the only people there, a few of the training school, the fire safety guys and some of the guys that were over in the workshops, and we sort of congregated in the yard not knowing what to do.

---

[91] Ken Harper.

So anyway the next thing Bumper Young came over, the driving instructor, and he handed me a first aid kit, and said, 'It's time you were out there trying to give a bit of help, see if you can do anything'. Obviously he didn't know how bad the situation was. I don't think he would have sent us out if he had known. So we all walked out carrying this stuff, walked round into Oxford Street and it was just mayhem.

I saw things that day that I never ever thought I would have to see; it was raw. The bus station itself was still on fire; at that time there were no appliances there. So we were sort of wandering about in all the rubble. There was really nothing we could do; we had no fire fighting equipment with us at the time. Then an appliance did turn up, I think they were from Holywood or Bangor. They got water onto the building. I remember coming outside and walking with Davey Page. There was a red brick wall just before a set of railings and as we got to the red brick wall there was a guy lying with his back against the wall and I went over to him with Davey and I bent down to see if he was OK and just as I got down he reached out and he grabbed the hold of me and the first thing he said to me was, 'What time is it?' Those are words that I'll never forget. Just those four words, what time is it? I didn't know what time it was so I said, 'Well it's three o'clock', I just had a guess. He said, 'Oh dear I've missed my bus.' He hadn't realised what had happened to him. We sort of patched him up a bit and the ambulance guys came over, lifted him up, and we were able to walk him onto an ambulance. We walked on up to the railings where the explosion had gone off, just to the right of it, just where the door was, where the bus drivers went in at the end of the shift. If you can imagine, like an egg slicer, the soldiers who had got out of their Land Rover were walking up the side of the railings between the bomb and the railings, the bomb had gone off and they were blown through the railings. So they were just mashed as they went through this set of iron railings, and their bodies were lying on the other side of the railings. So we went in round there and we started picking bits up and we put them into bags. Then we came back outside and started to help tackle the fire inside the building itself.[92]

---

[92] Stanley Spray.

As the emergency services couldn't transport all of the casualties to hospital, due to the sheer number of dead and injured people, a first aid post was established in the engine room of Chichester Street fire station. Firefighters who were between incidents, including Ken and I, worked with the injured trying to stabilise them until they could be taken to hospital.

Despite the carnage on Bloody Friday, the shooting, bombing and burning continued and on 26 July there were two Troubles-related fires in the city. The first was in a rope manufacturer's premises in Great Patrick Street, where, at about 9.45am, three armed men entered the building, held the staff at gunpoint, ordered them into the cloakroom and sprinkled petrol around the premises. They told staff that they had left a bomb with an anti-handling device on it and they should give them two minutes to get clear before raising the alarm. The stock in the premises was pretty flammable and the fire quickly got hold of the three-storey building. It wasn't until nearly midday that it was brought under control. The second fire was at a shop in Castle Place. An anonymous telephone warning said that there was a bomb in the building and, as staff were evacuating, a fire started, presumably due to some form of incendiary device. Despite the bomb warning, which may have been given to stop firefighters fighting the fire, operations continued with jets into the building and from the top of a turntable ladder. The second and top floors of the building were badly damaged. During the evening of 27 July there were two incendiary attacks, causing thousands of pounds worth of damage. The first was in the Spinning Mill in Royal Avenue and the second in a fashion shop in Upper North Street. The Spinning Mill was badly damaged and smoke billowed through the street as the fire was fought using jets and a turntable ladder. At the height of the incident two employees were rescued from the roof of the building using the turntable ladder.

The 9 August saw the first anniversary of internment and large crowds gathered to protest in the city. As the relatively peaceful protests broke up the violence started and buses were hijacked and set on fire. The Brigade turned out to many calls that day, one of which was to a bus on fire on the Falls Road. As the fire engines approached a crowd emerged from behind the bus carrying broken paving stones. The drivers of the appliances turned to make a getaway down the Falls Road, but before they could properly turn around they were hit with a barrage of stones and bottles. Then the shooting started, there were two bomb explosions in the city and a shop on the Antrim Road was badly damaged by fire. The violence and mayhem continued for days.

On 18 August, a fire, thought to be malicious as two previous attempts had been made to set fire to the shop, occurred in a discount store in Upper North Street. It took firefighters with five appliances half an hour to bring the large fire under control as flames leapt into the air, aerosol cans exploded and a pall of black smoke hung high over the city centre. Firefighters had a lucky escape that day when, after they had extinguished a fire thought to have been started by children in a lock up garage in Ballybeen, they found five flare pistols, ten flares, ammunition and a number of cs gas cartridges. On 21 August, a blast incendiary, incorporating 60 lbs of explosives, detonated at Stuart K Henry's paint depot off the Lisburn Road causing a huge ball of flame to leap into the air at the front of the premises. After the blast the fireball soon subsided but the air was filled with choking fumes as the paint burned and several paint cans exploded in the heat.

A Legitimate Target

During the Troubles the Fire Brigade were considered a neutral organisation by most protagonists, indeed the Brigade worked hard to maintain its neutrality. However, on 10 October things seemed to change when the Provisional IRA, in a statement issued in Dublin, said that they would treat firefighters as members of the security forces if they continued to assist the army in recovering military vehicles damaged by land mines. We were quite surprised by the statement as the Brigade was not equipped for vehicle recovery and no one had ever heard of us doing so, certainly not in Belfast. The next day two fire engines from Ardoyne were turned out to a house on fire in Alliance Avenue, started by an incendiary device, and as they approached six shots were fired in the direction of the appliance. It wasn't certain that the shots were being fired at firefighters and there were a number of soldiers in the area at the time.

The situation became even more frightening when on 5 December a workman working on the roof of a house in Ardoyne was shot dead by the army who mistook him for a sniper. Ardoyne was an area that had many open coal fires at that time and chimneys would often go on fire. The tactics of the Brigade at this type of incident was to have a firefighter climb onto the roof and extinguish the fire with a hose reel. This, of course, would mean that if a firefighter on the roof of a house in Ardoyne was mistaken for a gunman they could be in grave danger.

On Monday 30 October, people going into the city centre found it enclosed by a permanent metal security barrier. For many years, all vehicles and pedestrians would be searched by the security forces before being allowed into the city centre, in an attempt to prevent the kind of bombing and burning that had been going on over the previous three years. The Brigade was concerned that this new barrier, with its limited access points, would increase the time it took fire engines to get to large parts of the city and make it more difficult for civilians to evacuate if necessary. The Chief Fire Officer, Robert Mitchell, warned of the fire and escape hazards of the new gated barriers, suggesting that gated areas have a staffed gate to allow for escape in emergencies. He also said that in such an emergency the Fire Brigade might have no alternative but to cut through the gates, which over the years we did on numerous occasions.

The Belfast Sense of Humour

On 2 December there was a 100 lb bomb at the corner of Ann Street and Victoria Street. It was a Saturday and I was on the pump in Chichester Street. The location was not far from the station and we heard the loud bang of a big explosion, just before we were turned out to an actuated 77. As we approached the incident there were reports that there were other bombs in the area. This was not the biggest explosion I had ever attended but it certainly caused a great deal of structural damage to the buildings it affected. In fact the damage was so extensive that it was not known if the bomb was in the Lite Bite cafe or the RAF recruitment centre beside it. On arrival, we could see that there were a number of people trapped under the rubble. Sixteen people were injured, a number of them seriously. The damage, dust and debris spread over a number of streets. Because the damage was so widespread the crew I was with were assigned to the buildings in Church Lane behind the initial bombsite in Victoria Street.

We clambered over the rubble to the rear of the bar where some shouting was coming from and there at the back of the bar was a man sitting on the toilet, buried up to his waist in bricks and mortar, but otherwise seemingly unharmed. He was still holding his newspaper opened at the racing pages and as we started to dig him out he said, 'You can't even have a fucking shite in this god forsaken country'. We managed to dig him from the rubble and hurriedly started to help him from the danger area without properly pulling his trousers up, when:

> Sid Pollock had arrived whenever we were coming out the front of the building with this guy with his trousers down and told us make him decent – pull up his trousers. On the way in we had already spotted another device in among the rubble, the discipline was so strict that we actually moved him back about four or five paces so we could get his trousers on. We then walked him on out, and then informed him (Sid Pollock) that there was a second device and he says, 'Right, evacuate', so we all evacuated.[93]

For some reason we all thought this was hilarious.

1973

IN 1973 VIOLENCE in the city diminished slightly, although there were still many days of multiple deaths, many from gunshots. The petrol bombing, burning, nail bombing, stone throwing and rioting continued. On 10 January, a minor success was achieved for gender equality, when it was announced that three women had been recruited to work in the fire control room in Belfast. In what could be considered somewhat sexist language today it was reported that:

> Behind every successful man they say is a woman. And the 300 firemen who cover the Belfast area are no exception. For controlling their every movement are a team of three girls who have replaced male operators at the city fire headquarters control room in Chichester Street. The girls are an innovation but they have already proved their worth.[94]

Brian Douglas: Death of a Firefighter

As we came in for our last night shift, on 7 February 1973, we knew that we were in for another busy and difficult night. A one day strike had been called by loyalists and that night there was rioting, looting, burning and attacks on police stations, in a night dubbed 'the battle of Belfast' by one local paper.[95] The trouble started when loyalist protesters clashed with the security forces outside both Willowfield and Donegall Pass police stations. About 150 protesters broke away from the main demonstration in the west

---

[93] Ken Harper.
[94] *Belfast Telegraph*.
[95] *Belfast Telegraph*.

of the city and started to throw petrol bombs into buildings. They started serious fires in a pub; an office block and a number of cars were also set alight. As the violence continued, shooting started in the Ardoyne, Oldpark and Sandy Row areas. A church on the Woodstock Road was burnt out and pubs and shops on the Ormeau Road, Donegall Pass and the Lower Lisburn Road were set on fire. We went from call to call, dealing with incident after incident, so many it was impossible to keep an accurate count. From time to time it was necessary for fire officers to negotiate with the rioters in order to gain access to incidents. Then the worst happened. We lost one of our own when firefighter Brian Douglas was shot dead while fighting a fire in Bradbury Place.

I remember Brian well. When we first joined the Brigade we were on the same watch, Red Watch in Central, and we belonged to a small group of younger firefighters. Brian was a really decent guy who didn't smoke or drink and he spent most of his time off duty looking after his blind father, his mother having died from cancer several years previously. He had always wanted to be a firefighter and was particularly keen to join the Belfast Fire Brigade. He had recently been transferred from Central to Whitla station:

> An extremely nice guy, the only thing he ever wanted to do was to join the Fire Brigade. In fact before he came in to the full-time Fire Brigade as a firefighter he had worked in the control room in Lisburn. He was his father's carer, his father was blind. That was quite sad.[96]

Brian was shot while fighting a fire in the Groovy Boutique in Bradbury Place; he died in an ambulance on his way to hospital:

> About quarter to nine at night we got a call to go to a fire at Sandy Row. We went along the Dublin Road. We got to Shaftsbury Square and a policeman stopped us to say be careful, there's shooting in the area. Now we had heard this sort of thing many times and while you took note of it, and you did sort of keep your head down so to speak, but we've heard this many times.
>
> We got out of our fire engine parked on the Lisburn Road. Brian was in Whitla Street's fire engine, there were fire engines covering different areas and we had back up from Whitla. Brian was

---

[96] Jim Hughes.

upright, simple, uncomplicated, a decent guy. The sort of guy that didn't deserve what he got by a long stretch. Jack Fell was our boss and Sydney Pollock was the Deputy (Chief Officer) at the time. Big Sydney as we called him was there, and we looked around the corner and we couldn't get at the fire because it was down the street maybe 30 yards.

So Big Sydney says, 'Look if we run up from the Lisburn Road and get a jet from the other corner we could put water diagonally on it.' The reality was it wouldn't have mattered if we had got water on; the building was going to burn. But in those days you attacked everything from the inside, so Brian must have been trying to get a length of hose run out. I ran over to him and said to him, 'Have you got a branch?' And he said, 'No,' and I said, 'Right we'll go and get a branch.' So when we came back we plugged the branch in. I was number one, he was number two, we were probably one foot apart, I'm on the right hand side of the dry line of hose and Brian was on the left hand side. Our plan was to run more or less straight up the middle of the Lisburn Road to cross the head of Sandy Row. So I turned round and said, 'Right, are you ready to go?' And he said, 'Yes I am, let's go!' And, as hard as we could run, we then ran more or less up the middle of the Lisburn Road. To the right was the fire, and as we came into view at the junction from Sandy Row a machine gun opened up on us, a two-second blast of machine gun fire. I could sense the bullets in front of me, it seemed as if the bullets were only six or eight feet in front of me. I seemed to go into slow motion.

As soon as I got under cover I remember saying to myself, 'Brian', and Brian then, really wasn't that far behind me. And he was holding his hands over his heart and he says, 'I've been hit'. So I could see him almost starting to stumble and I put my hands out and got my two arms under his armpits, and he went down onto his knees, and we put him on his back. The boys then all ran over to help, and we got him onto his back and we opened his tunic and there was a wee hole just above his heart, with a one-inch trickle of blood. Then the ambulance came and took Brian away.

We packed everything up fairly quickly, there was no ceremony, everything, hose, stuffed into the crew cab and got back and we knew our colleague had been shot, and it really wasn't that long afterwards it was confirmed he was dead. The Chief came in to formally tell us that Brian was dead and we had a minute's silence, a really strange night. We got called out again about an hour later; we were still on the run. Nowadays you'd be taken off the run and offered counselling. We had plenty of other calls; in fact we were on the go until 6.00am. It was a wild night.[97]

Five people were killed that night, including Brian. Brian was 23 at the time, a little older than Ken and me. His father Harry was told of his only child's death just hours after a bomb in a garage on the other side of the road had shattered the windows at the front of his house. He said that Brian realised that being a firefighter was dangerous and that he had told him a year previously that if he died he would like a firefighter's funeral. This wish was granted on 10 February when hundreds of us, including firefighters from the Republic of Ireland and Scotland, paraded behind a turntable ladder carrying Brian's coffin with his helmet, belt and axe on top of it, to Carnmoney Cemetery where he was interned.

The army said that 30 or 40 automatic shots had been directed at the scene together with an RPG-7 rocket, which had missed its target and landed about 30 feet away.

Over the rest of the month there were shooting incidents in various parts of the city as the rioting, bombing and burning continued. For example, on the 12th, Red Watch's second day shift, we attended an explosion in a shoe shop in Royal Avenue. It was thought that two young women had planted the bomb and the resulting explosion caused a fire, which quickly raged through the four-storey building. Later that night a bomb was thrown into the Kingsway Bar from a passing car. The blast and the fire that followed reduced the bar to a pile of rubble. A shot was fired at the police and firefighters who attended the scene; it missed.

On 17 February, there was an unusual incident, even by Belfast standards. Two youths and a young woman, at least one of them armed, planted a bomb after holding up an assistant in a dry cleaners in Lower Donegall Street. The detonator exploded but did not fully ignite the incendiary materials that were attached to it. The small explosion damaged some containers of cleaning fluid, giving off clouds of highly toxic phosgene gas, the type of gas used in the trenches in the First World War. The Brigade loaned a breathing

---

[97] Ken Spence.

apparatus set to the ATO, who entered the premises to make the device safe, and firefighters went with him to disperse the gas with water spray.

During March there were many more bombs in the city, up to three or four on some days, and the shooting, rioting and petrol bombing continued throughout the month. On 12 March, four firefighters had a lucky escape when the first floor of a bombed building collapsed under them and sent them falling down to the ground floor. The incident started when two armed men planted two bombs in holdalls in the Buywell shop in North Street. The first holdall exploded and started a fire. Firefighters on the scene moved in to fight the fire despite fears that one of the bombs may not have detonated. The fire had been brought under control and four firefighters were on the first floor damping down when the floor collapsed.

At lunchtime on 16 March there were three explosions in the city. Two were in Smithfield, and the third was at a chemical merchant in Nelson Street. Armed men planted the two Smithfield bombs, one in a glazing shop and the other in a tobacconist's. The one in the glaziers had an incendiary device attached and when it detonated it started a large fire, which exploded out of the shop front, threatening the fire engine parked opposite. Firefighters moved in to fight the fire and, as they were preventing it from spreading to an adjacent paint shop, the second blast ripped through the tobacconist's just a few hundred yards away. An hour later a third bomb detonated causing structural damage to the wholesale chemists. Due to the chemical fumes given off at this incident, firefighters used breathing apparatus to enter the building to check it for fire and make it safe.

As the bombing and burning continued, a new weapon in the armoury of the ATO was seen in the city. A robot equipped with a camera, which allowed the bomb to be examined from a screen in an armoured personnel carrier some 50 yards away. Even though these mechanical devices became much more sophisticated as the years went by, this early one was fairly effective in letting the ATO have a look at the device before he approached to deal with it. Although the practice later stopped, at that time the ATO would let us watch their operations.

The night of 3 August was one of incendiaries and fire-raising in the city. At least five different buildings were attacked and four of them were badly damaged by the fierce fires that followed. On 8 August, over that night and into the anniversary of internment the following day, buses and lorries were hijacked and set alight. Two hoax bombs in dustbins caused traffic chaos in the city, there were attacks on six police stations and heavy gunfire could be heard as bonfires blocked streets and a significant number of people took part in demonstrations. On 15 August, a 200 lb car bomb exploded without any warning at the Sportsman's Bar in York Street. The barman, who lost a foot in the explosion, said he had bent down behind the bar to change a beer keg when he saw a flash and the wall crumbled. Firefighters, assisted by police and soldiers, dug in the rubble for more than an hour searching for survivors. One man, who was leaving the bar at the time, was killed and 11 others were injured.[98]

On 22 August, a fire badly damaged two floors of W S Mercer's electrical wholesale warehouse on the Ormeau Road. On the same day a young couple were injured when a number of explosive incendiary devices detonated in a house in Elaine Street, followed by a sheet of flame that spread through the building. It was thought that the house was a so-called bomb-making factory and that the couple had been preparing the devices for use at the time. The couple were seriously injured: the man had a hand blown off and the woman suffered serious burns, they both later died. She was 19 and he was 29.[99] Almost 50 lb. of explosives and eight incendiary devices, watches, batteries, fuses and detonators were discovered in the house. On 24 August, a 100 lb device in a suitcase was left in a taxi company office in Queens Square. As the ATO was preparing to examine the device it detonated, demolishing the building and causing severe damage to other buildings in the vicinity, including one face of the Albert Clock. Debris was thrown for hundreds of yards by the explosion and a huge pall of black smoke hung over the city as we fought the resulting fire.

On 25 August, three men were killed in a gun and bomb attack on a garage on the Cliftonville Road. As a result of the subsequent explosion a fierce fire broke out. Despite the danger from exploding paint tins and the possibility of further devices, firefighters broke into the garage and, using jets of water, they forced their way into the area where the men were thought to be, in a vain attempt to rescue them. All three of the men, the two brothers who owned the shop and their apprentice, had what appeared to be gunshot wounds.[100]

The rioting, shooting and petrol bombing continued and on 6 September a large modern office block in York Street was completely destroyed by a

---

[98] McKittrick, Kelters, Feeney and Thornton (1999).

[99] McKittrick, Kelters, Feeney and Thornton (1999).

[100] McKittrick, Kelters, Feeney and Thornton (1999).

bomb in a gas heater, left by three youths. When the device detonated in a ball of flame, the wall of the building was blown across the street and gas cylinders in the premises began to explode. Within minutes the whole building, which also housed an electrical shop and an army information bureau, was engulfed in flames. Despite falling masonry, the Brigade was able to use jets to extinguish the blaze. Crowds of people watched the firefighting efforts as the gas cylinders exploded, some of them making a louder noise than the bomb itself. With each explosion debris was thrown across the road and at the height of the fire a large pall of smoke hung over the street, obscuring the full extent of the damage.

Amalgamation: The End of the Belfast Fire Brigade

Even as the Troubles worsened the machinery of government continued to grind on and in 1970 changes to local government were being discussed. In September 1971 it was announced that the Belfast and Northern Ireland Fire Authorities would merge into a new unified Fire Authority for Northern Ireland in 1973. The new board would have 17 members, nine nominated by the Ministry of Home Affairs, four by the Belfast Corporation and four by an association of district councils. The authority was to elect its own chairman and members would not be paid a salary but would receive expenses. An order to create a single fire authority for Northern Ireland was moved in the House of Commons and approved by parliament on 5 February 1973, with the new Fire Authority for Northern Ireland coming into being on 1 October 1973.

On 23 March, in the run up to the amalgamation, Chief Officer Bob Mitchell announced his retirement following 11 years as Chief of the Belfast Fire Brigade. As amalgamation was so close it was decided not to appoint a permanent replacement and Sydney Pollock, the 'second officer', was appointed Acting Chief. Even though he was a stickler for discipline and could be hard on his firefighters, Big Sid, as he was known in the Brigade at the time, was a well-respected firefighter's officer. He would often be seen at major incidents leading from the front and encouraging his firefighters, particularly during the early years of the Troubles. In July George Morrison, Fire Force Commander of the Northern Ireland Fire Authority, was announced as the Chief Fire Officer for the new Brigade. George was a third generation firefighter with long standing connections with the Belfast Fire

Brigade, as his father and grandfather had served in the city and he was born in Ardoyne Fire Station. Sydney Pollock was appointed as Deputy Chief.

There was a certain amount of suspicion in Belfast regarding the amalgamation and our concerns were heightened when the Northern Ireland Fire Authority took a decision not to pay the full pension to widows of two firefighters who died in a fire in 1971. The two firefighters, Leading Fireman Leonard McCartney and Fireman Andrew Wiley, were killed while fighting a fire at the Melville Hotel in Londonderry on 21 November 1971. The Fire Authority decided to pay the widow's pension of £10.37 per week, but not the enhanced pension that we thought they should receive. The Fireman's Pension Scheme indicated that the widows concerned would be entitled to an augmented pension if the fatal injuries were received while saving lives, or, failing that, the injury was received 'otherwise than as aforesaid but in the course of duties performed'. It seemed clear cut to us, because the explanatory note stated that the Fire Authority could pay the enhanced pension if they thought it 'unreasonable not to do so'. The argument seemed to hinge on the authority's interpretation of the word 'otherwise' in the Fireman's Pension Scheme.

So on 30 August 1973 we took to the streets, trying to get public support and signatures on a petition for our campaign to get the Fire Authority to change its mind and use its discretionary powers to grant an enhanced pension to the widows of our fallen comrades. On the day, I found myself in a group that included Sammy Gamble and his two-year-old son at the front of the City Hall, and we had no problem at all getting people to sign the petition in support of our campaign. That evening's *Belfast Telegraph* had a picture of Sammy and his son on the front page holding a banner that said: 'When we give our lives take care of our wives'. The Fire Brigades Union presented a petition with 100,000 signatures to the Authority at a meeting in the Brigade headquarters building in Lisburn, as firefighters in Belfast and other parts of Northern Ireland continued collecting signatures on the streets. Eventually the Authority revoked its earlier decision and the full pension was paid. Mrs Mary McCartney, one of the widows concerned, thanked those who had campaigned on her behalf saying that, 'her life would be a little easier now'.[101]

Although the political decision had been made and amalgamation was by now bound to happen, the firefighters of Belfast, being proud of their

---

[101] *Belfast Telegraph*.

heritage, could not find it in their hearts to support the change. The union had to work hard to convince its members that the amalgamation would be a good thing:

> It was quite contentious. It got quite political at times and I know the Union were deeply involved. They had to persuade their own membership, in many cases, that amalgamation was good.[102]

And, when it did happen there was a certain amount of bad feeling at what was perceived to be a takeover by a predominantly retained service:

> I can certainly remember a lot of bad feeling at the time. There was the old pride of being a full-time service, highly disciplined, being taken over by a part-time fire service.[103]

Initially it seemed as if we had a point as the uniform, fire engines and stations all seemed to deteriorate fairly quickly after the event. One anecdote illustrates this point quite well:

> After being in the training centre, because of my size, they hadn't got a uniform for me. So although I was operational they couldn't get me boots, because I took size 13 at the time, they also hadn't got a tunic and it was the old single leggings. So I got all that and I got an old tunic, belonging to Sydney Pollock. He gave me one of his old tunics to do me until they got me one.[104]

Even the new fire authority recognised the difficulties of bringing two quite different organisations together:

> Any merger is fraught with difficulties at the outset when rumour is rife and fears abound. In these respects the fire service could claim no immunity, especially as one Brigade was comprised almost entirely of part-time firemen while the other was made up wholly of full-time personnel. It was only natural for existing

---

[102] Jim Hughes.
[103] Ken Harper.
[104] Joe Sloan.

staffs to be concerned about the effects of changes on their future careers.[105]

All in all it was a time of considerable upheaval when both services had their hands full with the operational requirements of the times. However, over time people's perceptions changed, and, even though pride in the old Belfast Fire Brigade remained as strong as it ever had been, it seemed that the general consensus was that in the longer term the amalgamation was a good thing for both services.

> Amalgamation was certainly good for firefighters, control staff etc. from the old Fire Authority for Northern Ireland who weren't on the same NJC conditions and all sorts of things, it was certainly good for them.[106]

Some people transferred from the old fire authority into Belfast and saw the transition from both sides:

> I was in the Fire Authority since '71 and came into Belfast in '73 so I sort of saw the amalgamation from both ends. I didn't find it difficult coming into the Belfast area because I was from Belfast, so I sort of knew my way around and I fitted in, maybe better than somebody from up the country, although there were a lot of guys from the country in the Belfast Fire Brigade. There was this old country cousin thing and the takeover and all that sort of stuff, which I think was actually probably justified. My view was, when I got there, that these guys knew their job, they were proud of what they did, for at that point over a hundred years, or damned close to it. These guys could do their job, as you know. OK, there were a few characters but by and large if you went out to a fire the fire went out, they knew what they were doing.[107]

A new Whitla Street station

On 18 September, just before amalgamation, the Lord Mayor of Belfast opened the new Whitla Street Fire Station, the last station commissioned

---

[105] Annual Report for the Fire Authority for Northern Ireland.

[106] Jim Hughes.

[107] Harry Welsh.

by the Belfast Corporation. The new station replaced one built in 1905 and it has a memorial to Brian Douglas in the form of a plaque on the door to the quiet room, which is dedicated to him. Brian's dad, who was at the ceremony, said:

> He would be very pleased if he knew his comrades had remembered him this way.[108]

A potentially embarrassing incident occurred during the ceremony when the Lord Mayor was trapped at the top of the Brigade's hydraulic platform, the only one in Ireland at the time, in front of the assembled guests. Eventually, after what seemed like a long wait, a control valve was bled to bring the platform down and allow the Lord Mayor to continue with the official opening. An experienced politician, he seemed to take this difficulty in his stride.

On the streets, the Troubles continued and the Brigade were kept busy with many incidents throughout October, November and December.

Car Bombs

In the early years of the Troubles bomb makers were restricted in their ability to construct large devices, due to the difficulty in obtaining commercial gelignite, their preferred explosive material. This changed over the course of 1972 and into 1973 when the car bomb asserted itself as the IRA's weapon of choice. By using a car the bomb makers could easily deliver a device to their target area, where it could fit into its surroundings until it detonated, often without warning.[109]

The tactics of bombing in the city also changed when bomb makers developed an ability to use widely available agricultural chemicals to make massive car bombs. Once they found that certain types of fertiliser and weedkiller, banned in Northern Ireland but legal in the Republic and England, could make very effective improvised explosive devices (IED's) there was virtually no restriction to the size or numbers of bomb they could make. The only major problem was that there was no really safe way to make this type of device and therefore many bomb makers died when the bomb they were working on, or transporting, detonated prematurely.

---

[108] *Belfast Telegraph*.
[109] Smith (2006).

The destructive effects of these tactics were seen during a seven-day period in November. On 12 November there were seven no-warning explosions in the city. About 13 people were injured and the morning rush hour was thrown into chaos as six of the bombs detonated within a two-hour period. Almost 1,000 lbs of chemical-based explosives were used. There was a 100 lb. car bomb at the Shaftsbury Bar on the Antrim Road, a bomb at the Mayfair Bar on the Grosvenor Road, a car bomb at the former SDLP offices in Killen Street, a car bomb at the American Bar in Princes Dock Street, a car bomb outside Farrell's Bar in Essex Street, a beer keg bomb in a car at a petrol station on the Oldpark Road, which was defused, and a car bomb at the Transport Bar on the Grosvenor Road. That night there was a car bomb in Ardoyne, a gas cylinder bomb in South Parade, a 100 to 150 lb. bomb outside a bookies shop on the Crumlin Road and a bomb outside a house on the Ormeau Road.

On 13 November, a bomb in a gas cylinder caused extensive damage to a bar on the Somerton Road when it was detonated by the ATO. On 14 November a 150 lb car bomb destroyed the Chlorine Bar in Gresham Street and caused damage to other buildings in the Smithfield area. The bar had just been repaired after an earlier explosion. A 200 lb car bomb exploded outside the Elbow Room on the Dublin Road causing considerable damage to the building and the surrounding area. This was followed by an explosion, which severely damaged the Albertbridge Wine Store in Lisbon Street. None of the devices had warnings given prior to their detonation, but they were all spotted before they went off and for that reason there were no casualties. On 15 November, three people were hurt when a car bomb detonated outside a pigeon club on the Falls Road, the Four in Hand Bar on the Lisburn Road was slightly damaged by a no warning explosion and two bombs, one of 100 lbs in a gas cylinder and one of 15 lbs in a beer keg left on a milk float in Brompton Park, were defused by the ATO. On 16 November, the Cherrymount Inn on the Crumlin Road was badly damaged by a no-warning car bomb and a bomb partially exploded in a bar in the Queen's Arcade where a police officer was slightly injured.

1974

ON NEW YEAR'S DAY 1974 a new power-sharing executive took office as a result of the Sunningdale Agreement. However, there was resistance from both the IRA and the UDA, and in May the Ulster Workers Council called a 'full scale constitutional stoppage', as power workers ran down electricity generation and supplies of food and petrol dwindled. After two weeks the devolved government collapsed. For the firefighters of Belfast the city was divided organisationally into two divisions as a result of the amalgamation and there were retained stations included in each division. The burning, bombing, rioting and shooting continued.

The Athletic Stores

On 23 January a number of incendiary devices caused a massive fire, which swept through the Athletic Stores building in Wellington Place, a traditional older building in the city centre housing a number of businesses. I was on duty and quickly found myself fully engaged in fighting the fire. The fire resisted our efforts and got a strong hold on the building by flashing up a lift shaft to the upper storeys. By that time we had managed to get three or maybe four jets of water onto the fire and a number of firefighters, including myself, stood in the street directing those jets into the building, as smoke and water spray whirled around us in the wind.

It was really cold but we were working hard enough to keep warm despite being soaked through. Jack Fell, our Station Officer, had gone into the building to have a look at how the fire was spreading and I have to admit to being a little surprised to see him stick his head out of a first floor window, upwind of the fire, to grab a breath and shout down for someone to radio Control and ask for assistance in a 'make pumps four' message. Leaving some jets working from the street a number of us then started to work our jets into the building to fight the fire in a close attack. We did eventually bring it under control without it spreading to adjacent property, but it took 40 firefighters, working with eight pumping appliances, two turntable ladders and a hydraulic platform.

During February, as the rioting continued, there were a number of shooting incidents, blast bombs were thrown, a number of lorries were set on fire and there were many bombings and fires to deal with. On the night of 19 February, a van bomb of between 200 and 500 lbs of explosives, in Castle Street, rocked the entire city centre when it detonated, causing considerable damage over a wide area and injuring five police officers and a firefighter. On 21 February, a bomb of about 30 lbs detonated in the Spa Inn Bar at the corner of Spamount Street and Trainfield Street just before midday. One man was killed and seven people were injured. Local people provided a ladder to allow a doctor and a nurse to assist a barmaid who was trapped by debris on the first floor. On their arrival, firefighters carried the woman out of the premises on a stretcher and searched the rubble on both floors for casualties.

At about 4.20pm on the afternoon of 22 February, there were nine incendiary devices planted in Woolworth's on the High Street. The area had been cleared when the first device detonated sending flames shooting through the four-storey complex. Firefighters who had been standing by got to work laying out jets of water to fight the fire. As the turntable ladder was being deployed there was a second muffled explosion, which rocked the ladder, and the firefighters who were operating it, backwards. There was then a third, fourth and fifth explosion as Woolworth's was engulfed in a wall of flame. In all 70 firefighters using ten appliances risked the explosions from blast incendiaries as they fought the fire; five appliances, including the turntable ladder, from the High Street side and five more from Ann Street on the opposite side of the block. Firefighters were further hampered as three more devices detonated inside the building over the next hour, shooting flames across the street and making firefighting operations difficult. Despite these difficulties, and the danger of further devices, fire damage was confined to the upper floors of the store, with the ground floor sustaining water damage. Once the fire had been extinguished firefighters stood by as the ATO checked for further devices and eventually cleared the building.

Overnight on 4 March a major fire was started by incendiaries in a printing works in Alfred Street. The building was well alight when the Brigade arrived and firefighters used water from eight jets and two height appliances to extinguish the blaze. Another malicious fire was started in a factory off Corporation Square on the same night. Huge clouds of smoke billowed over the city.

A Busy Day

Often during the Troubles there were very busy days for the fire service, they were frequently linked to events such as internment or the death of a hunger striker, or their anniversaries. However there were also many busy days, which were not connected to any specific event, so many that being exceptionally busy on any given day became almost normal for the Brigade.

15 March was such a day and during that afternoon there were 25 bomb warnings in the space of ten minutes, bringing the city to a standstill and severely stretching the Brigade. The first incendiary device actuated when it was carried by the security forces from Baird's shop in Royal Avenue, to waste ground beside the College of Art. Five people were treated for cuts and shock after an explosion in a shop in Donegall Square West. An incendiary device detonated in a shoe shop in Lower North Street and several incendiary devices were found inside the security zone. Then, a device detonated in White's Tavern at Winecellar Entry off Lombard Street causing an intense fire on the first floor. Ken McClune and I were on the pump that was dispatched to deal with the fire. In normal times this incident would have attracted several fire engines but on that day only one could be mobilised, because the Brigade was so busy. Due to a telephoned warning we arrived before the device detonated and started to set up, getting water from a hydrant into the pump and flaking out hose so that we could quickly get a jet onto any fire that was started by the device. We were made aware that there could be a number of devices in the building. As we were getting ready there was the unmistakable sound of an explosion coming from inside White's, and a fierce fire, obviously from an incendiary device, almost immediately engulfed the first floor of the building.

Ken and I quickly charged our hose with water, ran down to the bar and started to get water onto the fire through the first floor windows. George Roundtree was the leading firefighter in charge and he and the other firefighter from the pump managed to get a second jet onto the incident. Eventually we knocked the fire down enough to be able to get through the smoke and steam into the bar, so we could fully extinguish the fire. Every firefighter in Belfast was busy that day and help was drafted in from stations around the city. Things were so busy that the *News Letter* described the city as being in a 'virtual state of siege'.

On the morning of 23 March, together with other firefighters from Chichester Street, I attended a fire caused by an incendiary device in Easons Book Shop on Ann Street. We got the call after three devices had been found in the shop and were standing by when the first device detonated, starting a small but fierce fire. Despite our efforts at limiting the damage, Easons' second, third and fourth floors were badly damaged by fire only three nights later, following another incendiary attack. On the afternoon of 23 March, we attended a fire in North Street Arcade, which was started by an incendiary device. There was a warning that there were five devices in the building, but despite this we moved in to fight the fire. Several more incendiary devices, of the cassette type, actuated in a number of different premises in the arcade and this made firefighting operations quite difficult. We managed to extinguish all of the fires eventually, but by then several businesses had been badly damaged.

On the afternoon of 28 March, a 300 lb van bomb exploded at the army billet in the old Grand Central Hotel in Royal Avenue, causing a considerable amount of damage and starting a fierce fire. When the bomb detonated, the explosion was heard 20 miles away and buildings were shaken all over the city. A firefighter who was standing by at the time said:

> The lorry seemed to split in half and then I heard the tremendous bang. I ducked as pieces of debris came hurtling through the air.[110]

Firefighters moved in to bring the fire under control using four jets of water, but reports of a second device caused the withdrawal of the Brigade for a time. Despite a warning being given there were still several injuries caused by the explosion, and debris from the blast was scattered over the length of the street.

On the afternoon of 9 April a major assault was launched on the city when two car bombs and several incendiary devices detonated in and around the city centre. Gangs of youths went on a hijacking spree and in the west of the city blocks of streets were jammed with hijacked vehicles, many of which were set on fire. The Brigade was stretched to the limit as virtually all of its resources in the city and surrounding areas were committed to the various incidents. By mid-afternoon firefighters were working hard at controlling several major fires and a large pall of thick, black smoke hung over the city. Several people were injured at various incidents and the city ground to a halt.

On 11 April, the *Belfast Telegraph* reported that 1,000 people had been killed in Northern Ireland since 1969 as a result of the Troubles. The list of names covered two full pages.

On 24 April, the issue of four commemorative stamps by the Post Office celebrated the 200th anniversary of the birth of the fire service in Belfast. The stamps and commemorative postcards were sold in aid of the Fire Service National Benevolent Fund (since re-named the Fire Fighters Charity).

---

[110] *Belfast Telegraph.*

The Rose & Crown

The 2 May was a night shift for me. That night I watched a priest give the last rights to a human torso lying on the pavement outside the Rose and Crown Bar on the Ormeau Road, following an explosion. A human torso looking for all the world like a big lump of scorched meat with bits of burnt clothing stuck to it; it was difficult to tell if it was a man or a woman. I also did my best to help and rescue the injured and carried some of them, and perhaps some of the dead, to ambulances on a stretcher.

The incident started at about 10.15pm when Chichester Street received a call to an actuated 77 at the Rose and Crown Bar on the Ormeau Road. When we got there a few minutes later we were met with a scene of terrible devastation. The front of the bar was blown out and there were several people trapped in the rubble, which was all that remained of the bar, as well as a torso on the ground outside. Moving into what was by then a well-practiced routine, we suppressed our natural feelings for those that were dead or dying, closed off our emotions and did our best to help those we could. I worked with Billy Little, digging people out with our bare hands, helping ambulance staff treat those that were less badly injured and transporting those who were more seriously hurt on stretchers to ambulances for onward transmission to hospital. Other firefighters were doing the same.

We worked at the scene for over two hours, organising and helping a human chain to move the rubble and wreckage from the bar, which had only recently reopened after an earlier bomb attack. The canister bomb, which caused the damage on this occasion, had been thrown into the bar where it exploded almost immediately. Six men died, five at the time of the explosion and one nine days later. Eleven others were injured, an elderly man lost a leg and another man lost an arm. It was reported that laughter had been heard from the getaway car as it left the scene.[111] A *News Letter* report indicated that four bodies and 'half of a fifth body' were taken to the city mortuary. An elderly man who was having a drink in the bar when the explosion occurred said:

> I was blown off my feet, and when I looked round I saw the place was full of people lying in the most gruesome positions imaginable. Some were terribly injured. One man's leg was

---

[111] *Belfast Telegraph.*

hanging by a thread. Some were moaning in pain and others were cursing the people who had bombed the place.[112]

Two people were jailed for the atrocity; they were both 16 at the time of the bombing.

Smithfield Market

Smithfield Market was a Belfast institution. Brian Keenan described it well in his book, *I'll Tell Me Ma*:

> An enclosed street market that was more like a corner of Casablanca. You could buy anything in Smithfield. All kinds of people from all over the city came here. It was a world unto itself. Packed to the rooftops with second-hand furniture, memorabilia, record and book stalls, antiques, curios and oddities of every description stood waiting in this treasure house.[113]

My friends and I would often go there to look for records and hang out on a Saturday afternoon. On the night of 6 May, the market, a maze of little shops run by traders and arranged in a square, was totally destroyed by fire, leaving just a smouldering ruin. The fire, started by incendiary devices, was discovered just after 3.00am and firefighters with eight appliances, using 12 jets of water, fought it. Despite their efforts the old buildings, mainly made of wood and holding a high fire load, were quickly well alight and the roof soon collapsed. The only thing the Brigade could do was to prevent the fire from spreading to other buildings in the square.

On the morning of 9 May, the installation of new security gates in the city had some tragic consequences when a mother and her two children were killed in a house fire in Keegan Street. Firefighters were delayed in getting to the fire by the locked security gates and by the time they arrived there was little they could do to save the casualties. They used breathing apparatus to reach them but the mother was found on the floor of one of the bedrooms with the two children lying on a bed in the same room.

---

[112] *News Letter.*

[113] Keenan (2009).

A Foaming Pig

One tactic the army used to try and reduce the damage caused by explosive devices was to cover the device with foam, generated from equipment inside one of their armoured personnel carriers (nicknamed a pig). This newly converted vehicle was quickly nicknamed a foaming pig. On 10th May, the ATO used one of these 'foaming pigs' to 'damp down' a van bomb in Ormeau Avenue. An army spokesman said that a large area was saved from damage when they covered the 100 to 200 lb. van bomb with foam from a personnel carrier, filling part of the street with ten feet of foam: 'it was the first time we have used the foam on a live bomb and it really proved its worth. Damage would have been much worse'.[114] However, this tactic wasn't always successful:

> The army went through a phase of trying to soften the blow of car bombs exploding by using a foaming pig. They used to turn up with this pig and make foam. I can recall them at the side of the old Co-op building; a car sitting and they decided that they would try to envelope this car with foam. Well, they lost control, they couldn't get the foam switched off and they just about filled the street with foam and it was enveloping the pig that the foam was coming out of, and then it enveloped their other vehicles. And we were, of course, down the street laughing and laughing at this.[115]

The Ulster Workers' Council Strike

The Ulster Workers' Council Strike started on 15 May and, on 16 May, all the buses in the city were taken off the streets after a spate of hijackings, and roadblocks were implemented in loyalist areas of the city. At one stage there was panic buying in the shops and there were cuts in power supplies. My daughter Kerry was born just a couple of months before the strike and I remember queuing at a shop in May Street, which sold camping supplies, for three or four hours to buy a single ring, gas camping stove and a gas lamp so that we could heat her milk and feed her overnight. The shooting, rioting, street violence, petrol bombing, hijacking and setting fire to vehicles of all kinds continued until the 29 May when the recently devolved Northern

---

[114] *Belfast Telegraph.*
[115] Colin Lammy.

Ireland Government collapsed and the strike was called off. To a certain extent Brigade employees, being part of an emergency service, were given some leeway during the strike.

> There were all these roadblocks everywhere, and the paramilitaries would stop you, making sure you weren't breaking their strike. The worst part of it was getting to work because in work you were in the fire engine but going to work you were no different to anybody else.[116]

> I know the police and the army probably had a hard time but the ambulance service and ourselves, we were always let through, you showed your pass.[117]

There could still be difficulties of identification, however many firefighters who lived in or knew the areas affected would know some of the people involved in the barricades, and this could be advantageous.

> The barricades and the roadblocks were dreadful. We had one down at Holywood where I lived and the guy, he was supposed to be UDA, and they were all dressed up in their masks and all. There was one guy there who was actually a bin man and everybody knew him. I said, 'Geordie, will you get out of my road'.[118]

> There was a barricade at the front of Seymour Hill when I was going in to see my mum, and this guy, who I had known and grown up with, was standing there with a baseball bat asking me for my driving licence, asking me to get out of the car for his mate to frisk me and it was more than I could take at that time and I told him that I knew who he was, told him what I thought of him and my mother wasn't well and I was going in to see her and I wasn't going to be stopped by him. And I walked round and got into the car and drove off and they stopped me again coming out, to make life difficult for me. And I thought why did I do that? My mum and dad are living there.[119]

---

[116] Harry Welsh.

[117] Bob Pollock.

[118] Dessie McCullough.

[119] Ken Harper.

The bombing, burning, death and destruction continued throughout June and July and on 1 August a massive proxy bomb in a hijacked lorry exploded outside the Co-op Department Store in York Lane, badly damaging the building and setting it on fire. Friday 9 August was the anniversary of internment and this was marked by the now usual days and nights of shooting, stoning, rioting and burning.

An Unusual and Disturbing Fire

On 12 December one of the most unusual and disturbing fires I attended occurred in the Gate Lodge to the Queen's University of Belfast on the Malone Road. This was a small gatehouse in which two sisters aged 72 and 76 lived. The sisters were known as eccentrics in the city and they could often be seen collecting what could only be described as rubbish in the area around their house. The ground floor of the gatehouse was full of paper, bottles, clothing and old shoes, to a height of about five feet in one room. In the kitchen there were boxes and boxes of biscuits and breakfast cereals, bourbon creams seemed to be a particular favourite. Trails, where the rubbish only rose to about two feet, could be seen from the kitchen to the living room and rising to about a foot from the top of the door to the bedroom. It was thought that the two women slept on chairs in the living room or sacks on the rubbish in the bedroom. There was a space by the front door and another at the range in the kitchen. It appeared that the electricity had been disconnected some time previously and the sisters used candles to shed a little light in their sleeping and living areas.

That night I was one of the breathing apparatus wearers on the first attendance to reports of a fire in the gatehouse. When we arrived at the address we could see that there was a fire in the building and smoke was issuing from around the front door and window frames, but we could see little else. The density of the packed refuse had prevented oxygen from reaching the fire and for this reason it smouldered rather than bursting into flames.

We knew that the sisters were likely to be in the house so, after the front door was kicked in, my breathing apparatus partner and I were committed with a hose reel, on a search and rescue mission. Another two breathing apparatus wearers were quickly committed with us, and this meant that four of us were searching a room of eight feet by nine feet for two casualties, working in the rubbish by touch, due to the thick, black smoke. We couldn't find them! The fire did not take long to extinguish and the smoke and steam started to clear. Once this happened the sisters could be seen amid the rubbish in the room, one of them was face down and the other sitting in a chair. Both sisters had been overcome by smoke. We carried them out and they were put into the undertakers van to be taken for a post mortem.

1975

NINETEEN-SEVENTY-FIVE WAS SLIGHTLY quieter than previous years for the Brigade, even though the Troubles continued and there still were many major incidents and smaller fires to deal with. A decision was taken by the Fire Authority to close Ardoyne Fire Station and replace it with two new stations, one on the Springfield Road and one on the Westland Road.

On 21 January, after a quiet start to the year, bombers returned to the city centre when a car, which was moving in heavy traffic along Chichester Street near the city's main law courts and opposite Central Fire Station, suddenly exploded killing two men. The men died of multiple injuries and wreckage from the car was scattered over a wide area. A forensic scientist told the subsequent inquest that the bomb probably exploded as it was being primed.[120] Less than half an hour later another device caused blast damage to buildings and badly damaged a chemist shop in College Street. A fierce fire broke out engulfing four shops, when the bomb in a duffle bag detonated. On the 23rd, the Europa Hotel, marketed as the most bombed hotel in Northern Ireland (and eventually the world), was attacked again when an armed man planted a 30 lb suitcase bomb in the foyer. The device caused considerable damage to the foyer and broke many windows in the area.

On 26 January, a 16-year-old corporal in the Air Cadet Force was killed when a ten lb. booby-trap device exploded as he opened a door to the cadet clubroom in the grounds of Cavehill Primary School, on the Cavehill Road. Five other boys aged between 3 and 16 were injured. A neighbour described the scene:

> The ATC building was in a shambles. One wall was completely gone and wreckage was scattered over a wide area. In a corner I saw a body trapped underneath a pile of rubbish. It was a

---

[120] McKittrick, Kelters, Feeney and Thornton (1999).

horrible sight and obviously the lad was dead. His clothes were ripped to shreds. Blood was everywhere.[121]

The first fatality I got had a big impact on me because I wasn't that long in the job. It was over on the Old Cavehill Road. It was a portacabin for air training and one of the cadets had opened the door and it had been booby-trapped and he got blew away. I always remember that was the first I learnt about getting a bag to go and lift body parts. A few of us had to get what was left, it was really the legs and part of the torso, and put it on a stretcher to take it away. So that was the first big thing that has stuck with me.[122]

On 31 January, the city centre was thrown into chaos after two bombs exploded in clothes shops in Royal Avenue, with telephone warnings being given of more devices in the city. The first explosion was at Jackson's, the tailors. Two armed men planted the bomb. Three minutes after it detonated, the second device, at Temple's shop a few doors down in the same block, exploded, leading to speculation that the same men planted both bombs. A witness said at the time:

I was walking down the street when I heard the sirens and the Land Rovers pulled up outside Jackson's. They had just moved us back when there was a small explosion, which blew the front in. About five minutes later there was a much bigger explosion and black smoke poured out immediately.[123]

Another witness said:

There were flames coming from the roof too, and big clouds of black smoke came across the road.[124]

The traditionally built four-storey building was destroyed by the explosion and fire, and five shops were affected. The internal floors of the building collapsed and firefighters had to fight the fire while the ATO defused another

---

[121] McKittrick, Kelters, Feeney and Thornton (1999).

[122] Joe Sloan.

[123] *Belfast Telegraph*.

[124] *Belfast Telegraph*.

blast incendiary device, just 50 yards away. Hundreds of shops and offices were evacuated after the two explosions.

Bombs in Bars

The first weekend of April was a violent one. There were three bomb explosions, four kneecappings, 20 shootings and 11 people died. During the Saturday afternoon, McLaughlin's bar on the New Lodge was full of customers watching the Grand National on television, when a young woman left a cylinder bomb on the front step of the bar. The device exploded almost immediately killing two people. Seven others were injured, one very seriously. Just after 6.00pm that evening two youths walked into the Mountainview Tavern on the Shankill Road; one of them started shooting with an automatic pistol and as they left a box was placed just inside the door of the bar. A few seconds later there was an explosion and four people were killed by the blast, another died in hospital the next day. Sixty-one people were injured. A survivor said:

> I was hurled to the floor. I felt my face opening and a pain in my arm. I tried to get up, but I couldn't because of the blood and the beer on the floor.[125]

The Brigade attended with the pump and pump escape from Ardoyne and the emergency tender from Central, to try to rescue those trapped. Firefighters organised local people into a chain to move the rubble, while they worked inside the bar. One of those at the scene described seeing a woman with her clothes blown from her back, and her body pitted with lacerations from flying debris. The annual report of the Fire Authority indicated that 'this incident was typical of similar explosions and fire service personnel recovered the mutilated bodies under very harrowing conditions'. There were many explosive devices left in various public houses during the Troubles, often with fatal results.

The Bank Buildings

The Bank Buildings is one of the landmarks of Belfast. An imposing building dating from the 1780's it was constructed of red Dumfries sandstone on a

---

[125] McKittrick, Kelters, Feeney and Thornton (1999).

cast iron framework. As the name suggests it was originally a bank but had been converted into a department store in the 1850s. Even though it has an address in Castle Street, its polished marble entrance hall faces onto Royal Avenue. On 8 April there was a report that six devices had been planted in the building and four appliances from Central Station were mobilised to stand by.

> About three bombs went off and she went out through the roof in the Bank Buildings. Alex Aitkin and me got on the TL and I was about to get strapped in and Alex was sitting on it whenever, bang, just the next one went off. It just blew me down onto the ground; Alex he was blew off the seat. I remember George Morrison and Raymond Moore ran down Castle Street, hooked underneath my arms and brought me down into Donegall Place.[126]

The building was of five and six floors and about 250 feet by 850 feet. At the time it was being used as a department store and government offices. Three of the devices detonated and a fierce fire started on the upper floors. The first device exploded at 5.20pm in the restaurant on the second floor, and two others detonated within the next ten minutes. At the beginning of that year I had been promoted to Leading Firefighter and I was on the machine from Ardoyne that formed part of the make-up. By the time we arrived the fire had a firm hold on the old building and we could see that we were in for a difficult firefighting operation. However, we quickly got to work and I found myself in breathing apparatus, working a jet of water up the staircase to fight the fire on the upper floors. The fire was burning fiercely by that time, but we managed to get water onto it from the staircase while our colleagues did similarly, also from the inside, and other firefighters played jets in from the outside and from a turntable ladder. At one point I took up a position in the staircase with my right foot on a wooden chest, to help my BA partner and I balance the jet more comfortably. The fire took a number of hours to bring under control and over 30 hours to extinguish fully. In all, a total of 90 officers and men were engaged in fighting this particular fire and it was reported that 'the officers and men who first attended the incident were worthy of the highest praise for their courage and determination in the face of such a hazardous situation'.[127]

---

[126] Jimmy Armstrong.

[127] Annual Report for the Fire Authority for Northern Ireland.

It transpired that there was at least one other device in the building, and five days later demolition workers found a further bomb under debris on the third floor stairway. It was defused by the ATO.

Ken Harper remembers:

> Tommy Douglas said; 'Harper you remember you treated everybody to soup from the Salvation Army last night, up in the cafe?' I says, 'I do'. He says; 'well have a look at this'. He showed me a picture of a tea chest containing a device of nearly 500 lbs And he says; 'this was just beside the area where you were serving the soup'.[128]

It would seem that the wooden chest I had my foot on was the tea chest with the bomb in it.

At about 10.15pm on the evening of 12 April, four women and two men were killed and 29 people were injured when the Strand Bar in the Short Strand was attacked. It was a Saturday night and the bar was packed with drinkers, many of them elderly. Armed men entered the bar, sprayed the customers with machine gun fire and left a bomb. As they left they put a piece of wood through the handle of the door to stop the customers escaping.[129] The explosion occurred almost immediately, bringing down the roof of the bar and tons of debris. Firefighters from Central Station attended and, as there was no fire, they got to work immediately trying to rescue the people who were trapped. As was usual in this type of incident, it was impossible to keep concerned members of the public away from the scene, so firefighters attempted to put them to work clearing the rubble, while crews continued to try to rescue people at the locus. This incident was a particularly difficult one for the fire service to deal with because of the panic-stricken behaviour of many people in the area:[130]

> We could hear voices and I found a little crawl-way. I got the guys to tie a line round me and went in through this crawl-way. The only thing holding the bar up was the bar stools, but there was a poor woman trapped in it. I managed to get her hand and she kept saying to me; 'oh get me out; oh get me out, I'm dying'.

---

[128] Ken Harper.
[129] McKittrick, Kelters, Feeney and Thornton (1999).
[130] Annual Report for the Fire Authority for Northern Ireland.

But I couldn't do anything and she died holding my hand. I crawled back out of all this bloody mess and I was standing talking to Raymond Moore and Gerry and the gable wall fell and hit the three of us, and we ended up in hospital, a bloody gable wall fell down on top of us. It was actually Gerry O'Neil who pulled me away from it.[131]

Prisoners Escape

On 19 May, Central's location meant that it was involved in a slightly different incident when police opened fire on five remand prisoners, who were escaping from the magistrates' court next door. The prisoners concerned had prised open a skylight, got out onto the roof, scaled a barbed wire fence, climbed over the fire station wall, dropped into the station yard and ran through the side gate onto the street:

I was in Chichester Street when they broke out of the law courts; they landed on top of me. We were sitting at the wall, it was about lunchtime. We heard this commotion and this guy comes down and drops down on top of us, he just fell across me, he landed on top of us and took off across the yard. He was like greased lightning going out through the yard. How he found his way I don't know, he just went straight down the middle of the yard looked left and right, saw the open gate and just went for it. It was a wee bit of excitement; we all thought it was hilarious. As if we needed some excitement in them days.[132]

On the 30th a report into a conference organised by Belfast Women's Aid, heard that a doctor treated an average of one woman per day for injuries received at the hands of her husband. Firefighters can see the worst of humanity at times and I could relate that statistic to a fire I had attended about that time. The fire was in a terraced house in the west of the city and when we arrived we could see that it involved a mattress in one of the bedrooms upstairs. We took a hose reel up and quickly extinguished the fire. The room seemed a little strange as it had no furniture other than the mattress, and the only window was boarded up. There were a number of milk bottles filled with urine on the floor. It transpired that the man

---

[131] Ken McClune.

[132] Walter Mason.

who lived there kept his wife locked in the room for long periods and she had set fire to the mattress as a cry for help. George Marshall was the officer-in-charge and he informed social services as soon as we returned to the station. It seemed that was all we could do. From time to time I recall that incident, with not a little shame at sharing a gender with people who could do that kind of thing to another human being, and a society that allowed it to happen.

On 9 August, the anniversary of internment, barricades were erected on the Springfield Road, Falls Road and in New Barnsley. The crowds threw stones and blast bombs and several shots were fired. During the street violence a row of derelict houses and one occupied house were set on fire in Townsend Street at about 5.45pm. The rioting, shooting and burning continued over the following weekend. On the Sunday night a large fire broke out at Lowther and Harvey's hauliers' depot in Cupar Street, after a crowd broke into the premises. The firefighters who moved in to deal with the blaze were stoned by a mob, which hampered firefighting efforts, and the three-storey building was totally destroyed by fire. On the 13th four people were killed in a bomb and gun attack on the Bayardo Bar on the Shankill Road. About 60 people were injured, and one of them later died from her injuries. Two men were chatting at the doorway of the bar, one of them the doorman, when a green Audi car drew up and one of its occupants got out and opened fire, killing them both. Another man from the car then walked to the bar and left a duffle bag at the entrance. A woman later said that she saw smoke coming from the bag. As people were trying to get past the bomb it exploded killing two of them. The building collapsed, trapping many of the customers inside. Firefighters rushed to the scene, set up emergency lighting and pulled the victims from the rubble.

Rush Hour

Throughout the Troubles, bombers attempted to disrupt traffic during rush hour. This could be done through phoning in bomb scares or leaving elaborate hoaxes, which snarled up traffic and caused hours of delays. However, actual devices would often be used.

One example of this occurred on the 22nd when there were five explosions in the city. Four of them were during the busy evening rush hour and the resulting traffic chaos brought the city to a standstill. Two shops in Upper North Street, a garage and a shop in Great Victoria Street were the targets. In Upper North Street the bombs were left in bags in the doorways of McMurray's clothes shop and Smyth's record shop, shortly after 5.00pm. The bomb at McMurray's detonated 20 minutes after it was planted, badly damaging the shop. As the ATO started to work on the device at Smyth's, two more devices detonated in Great Victoria Street. The first was in the Stereo Sound music shop and the second in the next block at Prentice and Sons garage. The device at Stereo Sound detonated and the shop burst into flames. Despite the best efforts of firefighters the fire destroyed the shop, but it was prevented from spreading to the other shops in the building. Firefighting efforts were hampered by the bomb at Prentice and Sons, which also caused extensive damage and set two cars on fire, when it exploded 20 minutes after the device at Stereo Sound. The fire at Prentice and Sons was quickly brought under control by the Brigade. Meanwhile at Upper North Street the ATO was still examining the bomb that had been placed in the doorway of Smyth's record shop, and another ATO was examining a suspect device placed in public lavatories near the Brown Horse Bar. The ATO fired shots at the bomb in the doorway of the record shop in an attempt to defuse it but his efforts were unsuccessful and the bomb exploded just before 7.00pm. The shop burst into flames following the explosion and firefighters, who had been standing by in Royal Avenue, moved in to extinguish it. The premises were already beyond the point where they could be saved; however, firefighters prevented the fire from spreading throughout the building.

Victor McAllister

I learnt a great deal about firefighting and watch management on Red Watch Ardoyne. Jimmy Hamilton was the Sub Officer and George Marshall and Jimmy Douey were the other two Leading Firefighters. Victor McAllister was the Sub-Officer-In-charge. Victor was a very experienced firefighter and what he didn't know about firefighting or watch-management wasn't really worth knowing. He could be a bit grumpy at times but his real strength was his generosity to inexperienced junior officers like me and he taught me a great deal about the fire service, as well as preventing and sometimes covering my many mistakes. He was full of pithy sayings like, 'if you put enough water on it, it will go out' and 'we are not in a popularity contest you know'. People often talk glibly about heroes, but in fire service terms Victor was mine. He also inspired other firefighters: Charlie Hughes remembers that 'he was all round, he was brilliant on the fire-ground, and he was very, very good at man management'[133] and Kenny McLoughlin said, 'We wanted to impress Victor, because Victor was a good steady man'.[134]

Victor's skill and courage came to the fore, not for the first or last time, at an incident on the afternoon of 2 October when, having dealt with a number of bombing incidents earlier in the day, we got a call to a bomb-and-bullet attack on the Bush Bar in Leeson Street. A car had driven up; two men got out and fired shots into the bar, and then threw a bomb into it. Six people were injured in the resulting detonation. An 88 year-old woman lived next door, and on hearing the shots she started to go to her front door when the explosion demolished the wall between the bar and her house, trapping her under the rubble in the hallway.

The two appliances from Ardoyne arrived at the scene within a few minutes and we quickly organised a rescue operation. Working from the bar side, we located the woman's position and established that she was buried under the rubble. She was fully conscious in a gap under the debris and was able to communicate by shouting to us, although she seemed quite weak. We couldn't get to her from the bar side of the wreckage, due to the unstable nature of the debris, and so we started to plan a longer, more delicate rescue operation. The first thing to do was to gain access to her without making things worse, not an easy task in the circumstances. We found that she was up against the wall between her living room and her hallway, and that a door from the living room gave access to the hallway. More importantly there seemed to be a space along the wall through which we could tunnel towards her. Victor and I started to dig along the wall for the two or three yards that it would take to reach her. The tunnel we managed to make was about 18 inches or so high and wide enough to get her out. Victor took the lead and tunnelled through the space, passing back the bricks and masonry that was blocking our way, while I passed them out to other firefighters who were helping. I also tried to pack the rubble above us with the broken timbers, which had so recently formed part of the roof or ceiling above. In this way I tried to keep both Victor and the woman as safe as possible from the weight of the debris above us. Eventually we reached the woman, only to find that a heavy wooden joist had fallen on her foot and ankle, trapping her and bearing much of the weight of the rubble above. It was hot, dirty and dusty under there but she was alive, conscious and seemingly in good spirits,

---

[133] Charlie Hughes.
[134] Kenneth McLoughlin.

if firmly trapped. So there we were, we had reached her but we couldn't move her without risking bringing down the remains of the bar and part of her house down on top of us. What to do?

On the appliance we carried what is called a steel shod lever. This was a strong, approximately nine-foot-long length of four-by-four wood, steel shod and slightly angled at one end, which could be used to lever fairly heavy loads. Victor thought that as the woman was trapped next to the wall we could cut a hole in the brick wall beside her and use the lever to lift the joist off her ankle, and then drag her out of the tunnel we had made.

Victor decided that he would stay with her to give reassurance while other firefighters carried out the difficult task of knocking a hole in the wall, big enough to allow us to operate the lever. This was done with a cold chisel and a bolster hammer. I was given responsibility for health and safety (although we didn't call it that in those days) and communications, which basically meant sticking my head into the tunnel from time to time and shouting, 'are you all right in there Victor?' and then getting back out as quickly as possible. A number of bricks were removed with the bolster and chisel, but each time the chisel was hit with the bolster, a cloud of dust and small particles fell on the woman and Victor.

However, a large enough hole was eventually made and we could position the lever to try to move the joist, which was trapping our casualty. The lever was pushed into position and the weight of three or four firefighters was applied to it. The joist and the rubble pile creaked and for a second it looked as if we were going to fail. However, the joist moved the fraction of an inch it took for Victor to remove the woman's ankle from underneath it and drag her free. It took us most of the afternoon to get her out but the woman was alive, and we quickly got her onto an ambulance and off to hospital. The following year Victor would receive the British Empire Medal.

On the morning of 3 October one woman died, a second was badly injured and a young man was shot in the leg, when bombers attacked the East Belfast stationery shop of a prominent loyalist on the Albertbridge Road. A sister of the shop owner was with the woman who died when the bag was thrown in. She asked the youth at the door, 'what's this?' and he said, 'it's a bomb for you'. Other people who were in the shop managed to rush to the rear of the shop as the bomb exploded.[135] The frontage of the two-storey, traditionally built building was blown out, leaving a large pile of rubble on

---

[135] McKittrick, Kelters, Feeney and Thornton (1999).

the pavement. When firefighters arrived they frantically dug in the rubble to try to rescue the two women who had been trapped by falling masonry. The sister of the shop owner was seriously injured and was being helped from the rubble by firefighters when the dead woman's body was found. She had succumbed to head injuries caused by falling masonry and flying glass.

In November an on-going feud between the Provisional and the Official IRA worsened and there were several bombings, burnings and shootings. And, on 10 November, two appliances from Ardoyne were mobilised to a fire in the Gaelic Athletic Association Club in Beechfield Street. Victor McAllister was in charge and I was in the back of the pump escape. The club was holding a women-only 'mothers night' and no men were allowed on the premises, except the doorman. At about 9.30pm, three masked and armed men burst into the club and ordered the ten or so women who were there to line up facing the wall. The men sprinkled petrol on the floor and set it alight, leaving the premises as they did so. The women hearing the roar of flames discovered that the building was on fire when they turned round, and there was a wall of flame at about head height between them and the door. They had to jump through the fire to escape from the club. The fire badly damaged the club premises but no one was seriously injured.

The manager of the club was grateful enough to give Victor a case of beer 'for the boys to take back to the station' and Victor told me to put it on the back of the appliance. I was doing this when I was grabbed by the scruff of the neck and pushed against the wall outside by a man who I thought was the doorman. He looked massive to me and his nose seemed to be spread all over his face. 'What the fuck are you doing?' he said. I was trying to stutter a reply when Victor, who was coming out of the building behind me said: 'It's okay the guy inside gave it to us to take back to the station.' The doorman, who I was fairly sure was armed, put me down brushed me off and said: 'Sorry, I thought you were stealing that.' I have never been so glad to get away from an incident.

Firefighters are known for charitable fund-raising and those in Northern Ireland have always been involved in raising money for worthy causes. This continued throughout the Troubles and to give one example, on 15 November a 100-mile, four-day charity vintage pump pull through the Province in aid of the renal unit in the City Hospital was launched. One hundred firefighters were involved and they pulled the pump in teams of six.

1976

NINETEEN-SEVENTY-SIX WAS THE second-worst year for casualties during he Troubles. Political progress stalled and direct rule from Westminster continued. In September the scrapping of 'special category status' for prisoners came into force. The first IRA prisoner sent to the Maze prison after the cut-off date refused to wear uniform and wrapped himself in his cell blanket, the start of a protest that would escalate. The older stations in Belfast were in a poor condition but it was thought that the new station at Springfield Road should be completed during the year to replace Ardoyne, and work on the new station at Westland Road was well advanced.[136] Ardoyne was the busiest station in Northern Ireland and it was a great source of pride for those of us who served there that it would take two new stations to replace the old one. In operational terms the New Year began very much as the last one had ended and the bombing and burning continued throughout the year.

On 2 January, Northern Ireland was battered by hurricane force winds. This made firefighting operations very difficult when five business premises were destroyed in a half-hour series of bomb attacks that afternoon in Great Victoria Street, Howard Street and Donegall Street. The first two devices, of about five to ten lbs were planted by three armed men who held up staff at gunpoint on the ground floor of the Cavendish Woodhouse furniture shop, at the junction of Donegall Street and York Street. Three businesses were badly damaged but firefighters prevented the fire from spreading to the adjoining Co-op. Once the second device had detonated firefighters moved in and extinguished the fire fairly quickly. Just after the second detonation a third device exploded in a boutique in Howard Street, and almost immediately a fourth explosion occurred at Burrows Tailors in Great Victoria Street. Both of these explosions started large fires and huge palls of smoke could be seen over the city.

The North Street Shopping Arcade

Just before midday on 13 January, four people were killed and 20 more injured when an eight to ten lb. bomb detonated in a do-it-yourself shop within the North Street shopping arcade. Rescue efforts continued for well over an hour after the explosion and firefighters dug through the rubble in an attempt to rescue two people they knew to be buried under it. The

---

[136] Annual Report for the Fire Authority for Northern Ireland.

joint owner of the shop and a shop assistant were killed. The two other deaths were an 18-year-old woman and a 23-year-old man who had been assembling the bomb on the premises. A shop assistant, not the one to die, was said to have hidden the bomb in the false bottom of a hand cart the previous day and pushed it past security personnel to get it into the shop. The court was later told that the assistant had not returned to his job after the explosion and the police were unable to find him.[137]

> The whole front of the shop was blown out and there were wounded everywhere. In the actual shop itself, it was a mess. But on the stairs, as you came into the shop there was a set of stairs running up there, there was this torso sitting, no clothes or nothing on and you didn't know if it was a man or a woman or child or what it was, a torso was sitting there. And up the stairs was another part of a body lying just outside the toilet.[138]

> I was about a month or two into the job when we got the bomb in the North Street Arcade. There were four people killed. There was no fire as such; it was really sort of body recovery. It was mayhem. The bomb had gone off in the stairway and he was there, it was only his torso on the staircase.[139]

Just after midday on 20 January there was a major fire, once again driven by high winds, when two bombs exploded in Woolworth's in the High Street. The explosions, on the ground and first floors, caused a major fire to develop quickly inside the traditionally constructed building of four and five floors. As the high winds fanned the flames close attack firefighting was made difficult by the very thick smoke, which at one time totally blacked out the adjoining streets and rose high into the air over the city. A large crowd gathered to watch firefighters dealing with the blaze.

On 23 January, during disturbances in support of Frank Stagg, the IRA hunger striker held in an English prison, several vehicles were hijacked and set on fire in the city. As the demonstrations continued, I attended a bus alight at the top of the Whiterock Road. It was a one-pump attendance and George Marshall was in charge. When we arrived we saw a bus on fire and a large crowd standing nearby. As we gingerly prepared to fight the fire

[137] McKittrick, Kelters, Feeney and Thornton (1999).
[138] Brian Dynes.
[139] Kenneth McLaughlin.

George walked up to the crowd and spoke to the person who looked like he was in charge. The man said to George that he had a gun and that we were not to extinguish the fire on the bus.

We made up our gear and were preparing to leave when John Hyland arrived in his staff car. He asked George for an update and George told him what had been said, that the fire on the bus was not near any structures, would burn itself out and that we were leaving. John said, 'I know the area, let me go up and talk to him.' Both John and George started to walk up to the man and, being interested, I followed a few yards behind. As John approached, the man pulled a handgun from under his coat and said, 'Fuck off wee man.' 'Right,' said John, 'We're away.' And we went back to the station.

The Death of a Hunger Striker

On 12 February the hunger striker Frank Stagg died in prison in England. He was on his fourth hunger strike since being convicted on conspiracy charges in 1973, and his main demand was to be transferred to a prison in Northern Ireland. Following his death there was a series of bomb scares in the city; vehicles were hijacked and set on fire and there were several shooting incidents. That night was a violent one throughout Belfast, there were about 50 vehicles hijacked and burnt, blast bombs were thrown, there were also 17 shooting incidents and a dozen major fires. At one fire in Ardoyne, an angry crowd prevented firefighters from dealing with the incident and stones were thrown when the army arrived to secure the area. Bus services throughout the city were suspended for a number of days and a man died in a bomb explosion on the Falls Road. The following day the body of a teenager was found in the ruins of an Antrim Road furniture warehouse that had been petrol bombed and burnt out in the previous night's violence.

Reporters from both the BBC and ITV were with firefighters at this incident and they all scattered, as did the firefighters, when shots were fired in their direction. The violence continued throughout the month and on the afternoon of 16 February there were two further explosions in the city. The first, at the rear of the main post office building in Berry Street, started a fire. Then, some time later, an explosion designed to kill members of the emergency services who would have attended the first incident occurred at more or less the same place. This type of booby-trap was undoubtedly set up to kill and maim the soldiers or police officers that attended. However, they were an ever-present threat to firefighters, who would attend the same location to deal with the aftermath of the initial explosion. One of the ATOs remembers this incident.

> Control was tightened up and no one was to be allowed into the danger area. I was approached then by the Fire Brigade who wanted to enter the area to extinguish a fire that was threatening a 5,000-gallon tank of diesel used for refuelling GPO vehicles. They insisted and so I took the opportunity, with my No. 2, to go forward with them and reconnoitre the damage. The fire was extinguished and we all withdrew behind the cordon then and 30 minutes later a further small device detonated in the same building. Overall, a disaster had been averted simply by good luck more than anyone's good judgment.[140]

At about lunchtime on 17 February there was a massive fire caused by two bombs, which were planted by armed men in shops in Upper Donegall Street, the Co-op drapery store and Meenan's furniture shop. The device in the Co-op building detonated first, blowing the front of the building into the street and starting a fierce fire. Firefighters could not move in immediately due to fears of a second device, however a ground monitor was set up to play a jet of water into the building and this partially controlled the fire from the outside. Minutes later the bomb in Meenan's exploded and started another large fire, which quickly spread through the premises and threatened other businesses in the block. Flames leapt through the roof and smoke from the fire could be seen several miles away. The firefighting attack continued despite the suspicion of a third device in the premises.

On the night of 27 February loyalists took to the streets in a protest over the ending of special category detention for prisoners. Thirty-three vehicles were hijacked and 20 of them set on fire. Two bombs detonated; there were two shooting incidents, three petrol bomb attacks and two major fires in the city. Bus services were withdrawn after Citybus had five buses hijacked and set on fire. There were 54 bomb hoaxes causing traffic chaos and for the Brigade it was a hectic night. Crews attended 50 calls; half of them turned out to be fires and one of them was in a building on the site on Westland Road where the new fire station was being built.

On Saturday 28 February, it was announced that steel fencing and security gates would be provided to encircle the city centre (the so-called

---

[140] Wharton (2012).

'ring of steel') in an attempt to restrict the bombers ability to attack Belfast's main shopping area. That night, amidst the shooting incidents, dozens of vehicles were hijacked and set on fire, many malicious fires were started and the security forces were stretched by a series of bomb hoaxes. The biggest incident the Brigade dealt with was a major fire in a large single-storey storage depot in Cambrai Street just after 7.00pm. The premises were used for the storage of paint and carpets. The fire was eventually extinguished by 55 firefighters deploying 23 jets of water. In all 11 appliances were used from all five city stations, together with retained units from Glengormley and Carrickfergus.[141] The bombing and burning continued and, during the last weekend of the month, four soldiers were injured in a bomb attack on their observation post on the Oldpark Road.

Saturday 3 April was Grand National Day, and Daley's Bar on the Falls Road was full of people waiting to watch the big race when two men with guns got out of a passing car and threw a bomb into the bar. This was the third successive year that bars in the city had been attacked with bombs on the day of the Grand National, presumably because the bombers knew that the bars would be busy with customers. The explosion brought down one wall and the ceiling, and a second wall was left in a dangerous position. Injured people were taken to hospital in black taxis and ambulances as firefighters, who were quickly on the scene, dug people from the rubble and moved them to safety.

Just Another Day in the City

In the early hours of 5 April, Easons Bookshop in Oxford Street Bus Station was badly damaged by fire, and later that morning an armed gang planted a bomb in the Wellington Park Hotel on the Malone Road. This was the beginning of another day that put the Brigade under extreme pressure. Just after 2.00pm the device at the Wellington Park Hotel detonated and the resulting fire badly damaged the hotel. At about 2.30pm, a waitress carried a bomb out of the Milky Way Cafe in Royal Avenue as customers ran out of the building. The bomb later exploded but no one was hurt. Minutes later the biggest fire of the day was started by an explosion in the Frederick Thomas Pram and Toy Shop, also in Royal Avenue. A severe fire developed rapidly and, despite reports of other devices in both the toyshop and

---

[141] Annual Report for the Fire Authority for Northern Ireland.

premises nearby, firefighters mounted and maintained a firefighting attack, which confined the fire to one block of property. Ultimately the building was demolished on the advice of the building inspector from the city council.[142] Just before 3.00pm, the Headline Shipping Company's offices in Victoria Street were badly damaged by a bomb in a holdall, which was planted by two armed men. About the same time the first of three devices detonated in the Conway Hotel in Dunmurry and the resulting fire badly damaged a large part of the hotel. At about 4.45pm, a 15 lb. bomb was planted outside Steel's premises in Victoria Street and during that evening's rush hour there were a number of bomb scares that caused traffic chaos in the city. At 10.00pm, three petrol bombs were thrown into Mackie's factory on the Springfield Road but they were dealt with quickly and caused little damage. There were many days like this throughout the Troubles.

On 14 April, at least two devices were planted at the Linfield Industrial Estate. When they detonated they started a major fire, which quickly spread through the old mill and a number of adjoining premises. It was one of the biggest fires ever seen in the city and firefighting was made difficult due to high winds and the suspected presence of further devices. During operations, the roof of the mill collapsed and a massive pall of smoke hung over the city. It took four days for the fire to be completely extinguished.[143]

Linfield Mill was a cash and carry, it was a big store and it was a full attendance. There were a couple of bombs in it and I was in a crew with Bill Broadhurst, and we had worked our way up to the top floor. The place was burnt out and we had worked our way right up to the top. We took a jet with us and the building started vibrating, we had just gone in, hit over our heads with a jet. We thought oh! We've knocked something down, and just then Terry Dickson's head appeared at the window on top of the TL; 'boys did you not hear the whistles'. Apparently the vibration was the other side of the building collapsing. We were very lucky, but you know the funny bit, when he told us to get out we made up our jet and walked calmly down the stairs instead of getting out on the TL. I still laugh about that.[144]

---

[142] Annual Report for the Fire Authority for Northern Ireland.
[143] Annual Report for the Fire Authority for Northern Ireland.
[144] Walter Mason.

On 1 May a number of incendiary devices started a fire in a range of adjoining buildings, including offices, shops and a social club in Royal Avenue. The buildings and their contents were severely damaged by fire.[145]

During the late afternoon of 15 May, the Avenue Bar in Union Street was busy with customers watching an England v Scotland football match on the television, when a bomb in a duffle bag, filled with nuts and bolts, was thrown into the bar. The one lb. of explosives detonated and two men were killed. Twenty-seven people were taken to hospital and a one-year-old child in his mother's arms was injured in the street by flying debris. The ages of the injured ranged from one to 72. An angry crowd gathered in the area and made things difficult for firefighters, who were trying to rescue the injured. Eventually the dead and injured were removed from the wreckage and the firefighters withdrew.[146]

The Club Bar

On the night shift of 28 May, I was on out duty and in charge of the pump in Cadogan Fire Station, Davy Foy was in charge of the station. It was a Friday night and the bars in the university area were busy, with many students celebrating the end of their exams. The Club Bar on University Road was popular with both Catholic and Protestant students from nearby Queen's University, and it was packed with young happy drinkers. At about 10.15pm, we received a call for both of Cadogan's appliances to an explosion in the bar.

When we arrived a few minutes later we were met with a scene of utter devastation. Most of the students, many covered in blood and choking from the dust and smoke, had already scrambled through the front door, but quite a few more were still in the bar. The five lb. device had been left in the gent's toilets and, when it detonated, dozens of customers were showered with flying glass and pieces of masonry from a wall, which had been blown into the bar. The worst affected casualties were near the toilets and, when we arrived, a small group of them were still sitting at a table in that area of the bar. Above the table they were relatively undamaged, but under the table there was a tangle of legs and we were unsure which legs belonged to which torso. The first job was to untangle the legs and get the casualties onto stretchers. This we did, slowly and carefully, trying to prevent further injury.

[145] Annual Report for the Fire Authority for Northern Ireland.

[146] McKittrick, Kelters, Feeney and Thornton (1999).

Four ambulances then shuttled them to hospital. The people concerned were both male and female, some of them were quiet and unresponsive and some of them were screaming in pain. Two men died and 26 people were injured, five of them, including a 17-year-old girl, were seriously hurt and an elder brother of one of the dead men had to have both of his legs amputated below the knees. Once ambulance staff had taken the dead and wounded, we made sure the building was safe through the usual checks, structural stability, gas and electricity. We then helped the police gather the remaining body parts for forensic examination. One memory that stays with me is of finding a piece of a shinbone in the debris and putting it into an evidence bag.

On 8 June bombs badly damaged a plumber's merchant in West Street. Two firefighters were slightly injured when the second device detonated and they were hit with flying debris. On 14 June, two bombs caused extensive damage to a shop in Howard Street. The second device exploded about ten minutes after the first, while firefighters were in attendance. They remember that 'the place was going well... the sprinklers came on and managed to extinguish most of the fire. We were in the building when they found another device in it so it was everybody out again'.[147]

Two public houses, filled with lunchtime drinkers, were attacked with explosive devices on 17 June. Both bars, the Clorane and Madden's, were situated in the Smithfield area. A 30-minute warning was given and the Brigade was standing by when the bombs detonated. The explosions in both bars caused severe fires, and firefighters from all over the city were called in to fight them.

On 15 July a bomb explosion caused a major fire to develop in a four-storey range of city centre buildings. An armed gang of two men and a woman planted the device and it was thought they had left four devices in all. The first bomb detonated just before 11.30am and a column of flame and smoke could be seen rising from the building. Crews fought the fire from the street and worked hard to prevent its spread to other buildings nearby, spraying them with water.

A series of bomb attacks was carried out on 5 August with several devices detonating, some started fires that caused extensive damage to a number of buildings. At one of the incidents firefighters stood by on the Grosvenor Road as a major fire swept through the Calor Kozangas depot and gas cylinders started exploding, going off every few seconds. A large

---

[147] Gordon Mckee.

area was cordoned off and the fire was contained to the depot as firefighters prevented it spreading to adjoining property. A pall of thick, black smoke hung over the area as the explosions continued.

August 9 was the fifth anniversary of internment; it was marked by, the now usual, hijacking and burning in parts of West Belfast. This year the tension was heightened by the proposed removal of political status for Troubles-related prisoners. The violence started on the night of 8 August, continued until early next morning, went on all day and well into the night of 9 August. As the rioting continued dozens of vehicles were hijacked and burnt. On 10 August, the trouble intensified and there were gun battles and rioting as crowds of people set fire to a bus depot, a mill and a timber yard. Eight buses were destroyed in a petrol bomb attack on the Short Strand Citybus Depot and shots were fired as firefighters fought the fire. Petrol bombers attacked Corry's timber yard for the third time in a week and a gang of about a hundred people attacked Andrews Flour Mill, at the corner of the Falls Road and Percy Street. Petrol bombs were thrown into the building, which had also been set on fire earlier that week.

Petrol Bombs

Petrol bombs were often used as a weapon during the Troubles. They were easy to make and the materials they required could be easily obtained. They were often made more dangerous through the addition of a little sugar, which made them 'more sticky'. Early in the Troubles they were used in street rioting but soon they were being used to attack people in their homes.

On 27 August, a young couple and their baby died in their home as a result of a petrol bomb attack. A number of petrol bombs were thrown into the house by a gang of youths just before dawn. Firefighters fought the fire but were unable to save the family. The mother was 19, the father 22 and the baby was ten-months-old. At the inquest a woman said that she saw a youth breaking the front window of the house and another youth throwing in two objects. The flames quickly caught hold and after the youths broke the window of a house next door they ran off. One of them shouted: 'That's it come on'. A firefighter who was at the incident told the inquest he had used breathing apparatus to search the bedroom. He found the mother lying on the floor beside the bed and the baby was found nearby, in her father's arms. At the funeral the baby was carried in a small white coffin and hundreds of people, many of them women pushing prams, turned out to pay their respects.[148] A neighbour was badly injured while trying to rescue the family; he received severe burns to his back and was in hospital for six weeks.

In the early hours of 28 August, a brother and sister were injured when petrol bombers attacked their house in Gardner Street. The man involved had to jump from a first floor window to escape the flames. A firefighter was injured when he came into contact with an exposed electric wire while fighting the fire and he required hospital treatment. On 29 August, three petrol bombs were thrown into both the living room and bedroom of a house in Sicily Park. The house was badly damaged and a 22-year-old woman sustained severe burns to both of her legs. Her husband was slightly injured.

If a person was hit with a petrol bomb the results could be devastating. For example, in 1991, a bus driver was badly burnt when a petrol bomb was thrown into his vehicle. Several months later he still had difficulty opening his mouth, despite hours of operations, which were carried out by plastic surgeons in order to rebuild his face. Burned beyond recognition he suffered third degree burns to half of his body and speaking to the *Belfast Telegraph* several months after the incident he said:

> I remember this really bright flash and the bus just seemed to light up around me. At first I felt panic more than pain and I desperately tried to put the flames out and escape. My uniform just disintegrated and I could feel the body warmer coat I was wearing melting into my ribs. My hair, face, thighs, stomach, arms and hands were on fire. It was hell.

The bombing and burning continued to the end of the month and there were several major fires in various parts of the city.

The Belfast Telegraph Offices

On 15 September, at approximately 4.15pm, a stolen white Transit van pulled into the loading bay of the *Belfast Telegraph* offices in Royal Avenue. Two armed men jumped out, shouted a warning that there was a bomb in the van and opened fire on workers in the bay, wounding one man in the elbow. A soldier opened fire on them as they ran into Library Street but they made good their escape by breaking into a house in nearby Stephen Street. Their weapons were later found in the back yard of the house. The bomb,

---

[148] McKittrick, Kelters, Feeney and Thornton (1999).

estimated at between 50 and 100 lbs of commercial explosives detonated, badly damaging the structure of the building and starting a serious fire at the rear of the multi-storey property. One man was left badly injured and trapped under the debris following the explosion. Firefighters were quickly on the scene and started to tackle the fire while searching the building for casualties. The male employee was found lying among the debris on the first floor of the building by firefighters, who were working under very dangerous conditions and at great personal risk. He was quickly placed on a stretcher and taken to an ambulance, which took him to the Mater Hospital. His right leg was amputated just below the knee and he was placed in intensive care, however he died four days later as a result of his injuries.

On 17 September, firefighters attending a fire in a flat in Cliftonpark Avenue found the badly charred body of a 33-year-old woman beside her burning bed. There were three deliberately started fires in the house. As well as the fire involving the bed there were two more on the ground floor, one at the stove and another at the settee. A subsequent autopsy indicated that there was no soot or carbon monoxide in the body, indicating that the woman was dead before the fires were started. Tests indicated that she did not die from cyanide poisoning but unfortunately a fire in the forensic science laboratory destroyed the remaining tissues before further tests for other poisons could be carried out. The woman had a history of mental stress, was on a course of medication and was said to have kept herself to herself. While the cause of death remained unknown the police treated the incident as a murder.

On 18 September, an explosion started a fierce fire in the forensic science complex at Newtownbreda. The water mains for the premises were switched off, due to water conservation orders in force at the time, and this made firefighting operations more difficult than they needed to be. The fire involved a number of gas cylinders, which exploded in the heat and injured two firefighters. There was some confusion as to how the incident started, but it was thought to have been an accident.

Danger Money

As the Troubles progressed, organisations such as the police and prison service were granted a number of additional monetary allowances to make them more attractive to potential applicants, during what by then were very difficult times. Both of these professions were better paid than the fire service. Towards the end of 1975 and into 1976 the feeling grew among firefighters that their work was undervalued in comparison with other public services and something should be done about it: 'Part of the reason people wanted danger money was because everyone else seemed to get it', firefighter Roger Dawson recalls.[149]

As a result of this feeling within the workforce, it was announced on 3 July that firefighters would launch a campaign for substantial danger money. Negotiations were started with the Fire Authority, but after several months it seemed that no progress was being made and attitudes hardened. One of the difficulties for the authority at the time was that the Government had agreed a pay policy with the Trades Union Council, and to give the firefighters of Northern Ireland the kind of pay rise being talked about would have broken that policy. By the end of September 1976, some firefighters in Belfast had decided to take unofficial action in support of the claim, and it was reported in the local media that they were going to refuse to fight fires caused by terrorism unless they were paid danger money:

> We decided to form our own committee and try and organise something that would produce danger money of some shape or form, unofficially. The decision was that if we were going to call any industrial action it was only to be against commercial targets. Anything else we would turn out to.[150]

The Fire Brigades Union (FBU) did not support the action, stressing it was unofficial and that they were pursuing a claim for danger money through the usual channels:

> We tried to put it to the union a whole lot of times, but their opinion was that if they gave us danger money they would have to give Rotherham danger money and they would have to agree to London having danger money and it was too complicated.[151]

A 15-strong action committee was formed and it discussed how to carry out the boycott. At the same time attempts to persuade the retained not to come into the city to deal with such incidents were made, and seemed to be having some success. Following the committee meeting Brian Dynes, who became the committee's spokesperson, said the members had agreed that, where life

---

[149] Roger Dawson.
[150] Brian Dynes.
[151] Brian Dynes.

was in danger, appliances from Belfast would turn out to explosions. On the evening of 4 November it was agreed by the action committee that the boycott would be extended to cinemas, theatres, bars and clubs. As all this was going on there was just one civil disturbance call in the city, to the Casanova Restaurant, which the Belfast firefighters attended. For some reason the city was uncharacteristically quiet.

However, on the night of 13 November a five to ten lb. device started a fire in the National Cash Register building at Skegoneill Avenue. Four appliances attended but when they found that there was no risk to life the firefighters withdrew and let the building burn: 'The first we got during that strike was one night to Olivetti off Skegoneill Avenue, typewriters and stuff. They went on fire; we just stood back.'[152] The fire was eventually extinguished by retained crews from outside the city, but not before the premises were virtually destroyed.

The Gas Works

On 16 October three men entered the Belfast Gas Works on the Ormeau Road in an attempt to plant a number of bombs at an army post within the grounds. It would seem that one of the devices detonated prematurely, setting off four other explosions. The three men, who were near the security fence inside the complex, about 25 feet from the gasholder and 75 yards away from the army post, were killed. The explosion was so severe that body parts were found 150 yards away, and it was some time before it could be established how many men had died. A fire was started by the detonations and firefighters from the city were dealing with it when a nearby gasholder ignited and generated a huge fireball, which could be seen from ten miles away.[153] 'The work of some of the firefighters was outstanding and a major disaster was averted.'[154] Hundreds of people were evacuated from their homes in the area as firefighting operations continued:

> They were carrying a bomb round to blow the army up, a command post at the gas works. The bomb went off prematurely, killed the boys and blew up the gasometer. So we ended up there all night. But luckily the way that gasometer went up, it went up mostly

---

[152] Jimmy Armstrong.

[153] McKittrick, Kelters, Feeney and Thornton (1999).

[154] Annual Report for the Fire Authority for Northern Ireland.

into the air. If it went outwards, god knows what would have happened. I remember it burning and burning, it just got more and more intense and then the gas just erupted, the whole thing went up. All of Belfast was lit up for about 20 or 30 seconds.[155]

Even though the unofficial industrial action was going on at the time firefighters attended and dealt with the incident, and, on 19 October, the Fire Authority issued a statement, which read:

> The skill, courage and devotion to duty of the members of the Brigade who turned out were in the highest traditions of the fire service, and by their actions they undoubtedly prevented what could have been a major disaster. The statement added: At the same time we are saddened and disappointed by the decision of some firemen in Belfast to continue their unofficial boycott of fires caused by terrorist action in some classes of premises when lives are not at risk.[156]

That day two bombs detonated in adjoining shops on the Shankill Road, starting fires which were extinguished by retained crews from Castlereigh and Holywood because the city firefighters refused to deal with them, due to the campaign for danger money. At one of the incidents firefighters from Ardoyne did damp down the nearby Elim Pentecostal Church, which was threatened by the spreading flames, saying that they did 'run the hose out, but all we did was actually wash down the church next door to stop the fire. We surrounded the fire, and wouldn't let it affect the residential houses'.[157] However, by this time the inaction of the city firefighters was bringing condemnation from local people.

The Worst Incident

Although I suppose it shouldn't have, one of the things that surprised me when I was carrying out the research for this book was the frequency with which firefighters talked about incidents that had nothing to do with the Troubles. They often thought the worst incident they had attended was not Troubles-related, but usually included the death of civilians, often children.

---

[155] William McClay.

[156] *Belfast Telegraph.*

[157] Joe Sloan.

For me this included a fire in a terraced home in Matchett Street, just off the Shankill Road.

Friday 5 November was a night shift for the Red Watch and Steven Murray and I were rostered as the breathing apparatus wearers on the back of the pump escape. It had been a relatively quiet night when we received a person's reported call to a house on fire in Matchett Street at about 1.30am on the Saturday morning. As we turned into the street we could see fire coming from most of the front windows of the terraced house and we knew that this was going to be a difficult one. Victor McAllister was in charge and neighbours told him that they had heard screams for help; Donna, the eight-year-old granddaughter of the family, had been seen at the front window. As was the practice, Stephen and I were already dressed in breathing apparatus and Victor ordered us to lay out a jet of water and fight our way up the stairs, to try to rescue the family members we knew were in the building. We ran out the jet, kicked in the front door and, having sprayed our hose around the entrance and staircase to bring the temperature down a bit, we made our way up the stairs, extinguishing the fire as we went. The whole ground floor was well alight and other firefighters started to attack the fire in that area of the house with a second jet. The fire had a good hold upstairs, and the plaster fell off the walls as we made our way up the stairs, exposing some of the electrical wiring.

Having made it to the top of the stairs we could see three casualties in the back room. Eight-year-old Donna, Mary, her 51-year-old grandmother, and her 31-year-old aunt who was also called Mary; all three were badly burnt. We quickly hit the fire in the front room, to secure an exit back down the stairs, and went into the back room to start the rescue. We could see that all three casualties were in a bad way, their clothes had been burnt off and their bodies were burnt and swollen from the fire, however at least one of them was alive as we could hear groaning. We lifted the first casualty, with Stephen taking her legs and me lifting her under the armpits, and we moved her out of the bedroom and onto the landing. However, as we got to the top of the stairs, we had to stop as she was slipping from our grasp (in those days firefighters didn't wear gloves for this kind of work). Stephen was one or two steps down and I was at the head of the stairs. What we didn't know was that some of the electrical wires had come down and they were still live. When we picked the casualty up again we both received an electric shock. Stephen was thrown down the stairs and landed unconscious at the bottom. I was luckier; because I was at the top of the stairs I was only blown back into the wall. I then did the only thing I could to get the casualty out; I pushed her off the wire with my rubber boot, to avoid being electrocuted again, and shoved her down the stairs. I couldn't carry her in the circumstances due to the narrowness of the stairway and her position on the stairs. By then other firefighters were there to help and they carried her from the entrance of the house to the ambulance. Another crew had donned breathing apparatus and the three of us went back into the house and up the stairs to carry out the other two casualties, one of whom was eight-year-old Donna. I was fairly sure that at least two of the women were alive when we carried them out of the room, but sadly none of them survived. Stephen was in the ambulance when I came out of the house with the last of the casualties and Roger Gotch, who had by that time taken charge of the incident, indicated that I should go to hospital with him, even though I didn't feel I really needed to.

In the casualty department we were treated for cuts and electric shock while the nursing staff worked hard to save two of the badly burnt casualties. Not for the first or last time I was filled with admiration for the professionalism and skill of the trauma nurses as I watched them doing their best for the badly burnt women. It was later determined that the fire had almost certainly been started accidentally, by someone dropping a lit cigarette end on an easy chair in the front room. The grandfather, who was in the house at the time of the fire, told the inquest that he was asleep when his wife came running into the room screaming that the house was on fire. When he tried to rescue the others, who were also asleep, he was overcome by smoke, fell out of a window and was seriously injured. Verdicts of misadventure were returned on Mrs Mary Irvine, her daughter Miss Mary Elizabeth Irvine and Mrs Irvine's young granddaughter Donna Herd. The coroners court was told that Miss Mary Irvine was dead when brought out of the house by firefighters and the other two died later in hospital. I still wish we could have done more to save them.

The Unofficial Action Ends

The unofficial action was in its seventh week when there was a turning point; a bomb badly damaged the headquarters of the National Trust at Barnetts Park on 11 November. A group of men entered the building and made staff lie on the floor as they planted two devices. Retained firefighters with five appliances attended and fought the fire, which was started by the resulting explosions. Directed by staff, they entered the burning building to salvage valuable artefacts such as paintings, antique furniture, china and silverware, which they brought outside. However, irreplaceable records relating to the Trust and going back to the 1930s were destroyed. Two crews from Cadogan turned back from the incident when they learned that no lives were at risk. The next day the Fire Authority and the City Council started to take a tougher line with firefighters who were taking part in the unofficial action, threatening to sack them. The action continued, but it 'started to go downhill then and the public were turning against it'.[158]

On 15 November, a bomb planted by two armed men started a fire in Beggs and Partners in Little Patrick Street. In all, six full-time appliances were mobilised to this incident. Firefighters had an 'impromptu meeting on the street as each machine came up' and said 'we're not dealing with this, it's a civil disturbance call';[159] then they went back to their station.

However, by this time a number of firefighters thought that things had gone too far, and some of the officers in the city decided to try and do something about the incident.

Ken Harper remembers:

> It ended up with Jack Schofield, myself and Leslie Johnston, one on the hydrant and the other two of us on the jet, putting the fire out, while over 40 firefighters watched us and wouldn't tackle the fire. We got a lot of heckling from a lot of people and I thought that's desperately hurtful.[160]

But even with this intervention the building was well alight before part-time firefighters from two stations arrived to the cheers of members of the public. However, many firefighters were, by now, feeling conflicted:

> We got a call to a wee printing place and we arrived at it and Eddie Hay was the LF and I said to him, 'Eddie it's not that big a fire, we'll put it out,' he says, 'we can't, we can't, no.' I said, 'look if we put it out who's going to know?' And this guy was standing, he was near pleading with us like, he said, 'I have stuff in there, papers that I really, really need and there is a safe there

---

[158] Brian Dynes.

[159] Joe Sloan.

[160] Ken Harper.

and everything I have is in it,' and I said, 'look, give us the keys or come in with me'. So he gave me the keys of the safe and me and Eddie Hay went in and we opened the safe and emptied it out into a bag and all the papers I could see on his desk just pulled them all into the bag and went out. But his wee place burnt out. It was really awful. It really was.[161]

Members of the action committee were put under pressure and various threats were issued to its members, especially its spokesperson. By now firefighters had had enough, support had faded away and crews on the stations refused to carry on with the dispute; the action committee disbanded, saying that they were prepared to leave the matter in the hands of FBU officials and the Government, and that they were disappointed at the bad public feeling the dispute had created. It was back to business as usual and the bombing, burning and killing continued to the end of the year.

---

[161] Dessie McCullough.

Billy White, just after being petrol bombed on the Falls Road in 1969.

The Junior Firefighters squad of 1969. Brian Allaway is on the front row, fourth from the right.
© Brian Allaway

The early bombs were quite primitive, they used timers comprising clothes pegs held open by rubber bands with tintacks as contacts.

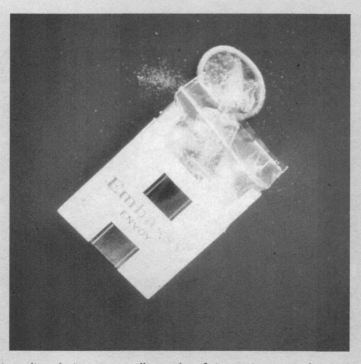

Some incendiary devices were small enough to fit into a cigarette packet. In this type, some acid was put into a condom and when it burnt through it would ignite the flammable material in the packet.

During the water shortages we used road tankers to take water for us to use at incidents. On one occasion we were quite amused that two of the tanks were labelled 'Guinness', if somewhat disappointed that they only contained water.

Some of the explosions seen in the city were extremely powerful. Cars and buildings could be thrown about as if they were toys.

White Watch Ardoyne, just before the move to the new station on the Springfield Road. Brian Allaway is in the middle of the front row. © Brian Allaway

The tax office in Belfast was a frequent target for the bombers.

Central's location meant that when the courts were attacked in terrorist operations, which were a frequent occurrence during the Troubles, the station often sustained collateral damage. On 29 July 1985, a van, which was packed with approximately 500 lbs of explosives, was left outside the Recorder's Court on Chichester Street. When the device detonated 30 minutes later, the blast could be heard ten miles away.

Belfast's Central Station on Chichester Street did have one disadvantage in the Northern Ireland context. It was right next door to the Petty Sessions Court, near to the Recorder's Court and directly opposite the High Court.

There were many spectacular fires in the city over the 25 years of the Troubles.

Shops were a favourite target for bombers and fire raisers, particularly the larger ones, and on these occasions roller shutter doors could either protect some of the premises or make it difficult for the brigade to gain access.

The Grand Opera House was badly damaged by explosive devices on more than one occasion during the Troubles. On 20 May 1993 it was severely damaged when a 1,000 lb explosive device, hidden under scrap metal in a skip lorry, detonated. Thirteen people were injured and damage was caused to many other buildings in the area.

Fire Control in the Headquarters building in Lisburn. During the Troubles, control were the unsung heroes.

On Friday 2 July 1993, there were numerous petrol bomb attacks in the city and more than 70 vehicles were hijacked and burnt. The rioters attacked the firefighters who were fighting the fires and, during the night, an appliance from Cadogan Fire Station was hijacked at gunpoint and burnt on the Donegall Road.

At lunchtime on 23 October 1993, a bomb was left in Frizzell's fishmongers when the shop was full of people on a busy Saturday afternoon. The device detonated almost immediately, killing ten people, including one of the men who had planted it. This incident had long lasting effects on the firefighters who attended. Note the precarious position of the roof while firefighters and others work below it.

PART THREE

1977 to 1984
A Plateau of Death and Destruction

1977

THERE WAS A marked drop in the overall death toll in 1977, as the intensity of previous years couldn't be sustained. However, the Troubles settled into a terrible continuum of death and destruction, which would continue for many years. Industrial relations worsened in the UK generally and at the end of the year the fire service began its first ever national strike over pay. While arguments were put forward that, because of the circumstances in Northern Ireland, the Brigade should not take part, full-time firefighters in Northern Ireland and some of their retained colleagues did go on strike. The dispute rumbled on over Christmas and into 1978. However, the issue over danger money in Northern Ireland was resolved when that dispute was taken to arbitration on 27 January. The tribunal found in favour of the FBU's claim that there had been a significant and identifiable increase in the responsibility of Northern Ireland's firefighters since 1971. However, with regard to amount, they awarded £2 a week, backdated to 18 October 1976, instead of the £5 a week that was claimed.

**

At approximately 1.30am on 21 January a number of incendiary devices detonated in Fraser's clothes shop in Castle Street. The building was filled with thick, black smoke when the Brigade arrived and, as it was known that there was a caretaker in the building, firefighters wearing breathing apparatus climbed the stairs and rescued him while others fought the fire. Firefighters administered mouth-to-mouth resuscitation as they brought the caretaker out but he was found to be dead at the scene. There were still a number of devices in the building and some of them detonated, making the fire burn with even more intensity.[162] On 23 January, the bodies of two men were found in a burning car in Downing Street. Both men had been stabbed and shot. This was the type of incident that firefighters had to deal with from time to time.

On the afternoon of 18 January, gangs of youths set fire to a number of buses in what seemed to be a concerted attack. Four buses were set on fire in the west of the city, a suspect proxy bomb was left in Glengall Street Bus Terminus and petrol bombs were thrown into Ardoyne Bus Terminus. On 19 January, attacks on buses continued when two were hijacked and set on fire on the Donegall Road, and another ten were destroyed when a gang broke into the Short Strand Depot and set them alight. The same night, arsonists broke into a youth club in Nubia Street and started three fires using petrol as an accelerant. The fires took hold and it took firefighters some time to bring them under control. On 26 January, a number of explosive devices were planted in a four-storey, traditionally built building in Adelaide Street, which was being used as a wholesale motor accessory suppliers. On 21 January, a major fire destroyed Gillespie and Wilson's furniture store on the Albertbridge Road. A butcher's shop and a vacant fruit store were badly damaged and a fish shop, an optician's, a shoe shop and a private medical practice were also damaged. Six people from a small block of flats lost their homes. It took firefighters with eight appliances several hours to bring the fire in the three-storey building under control and they were still damping down the following day. It was estimated that the damage would cost almost a million pounds to put right. At about 1.45pm on 24 January, a hijacked oil tanker carrying 2,600 gallons of fuel oil was driven into Donegall Pass with an explosive device suspended inside the tank. The following detonation caused a major fire, which was extinguished by 34 firefighters with seven appliances, using seven jets and two foam branches, despite the hazards posed by the difficult circumstances.[163] On 28 January, a similar device in a tanker in Dunmurry was defused by the ATO. In the early hours of 30 January, a carpet warehouse in Dunbar Street was destroyed by fire.

On Saturday 21 May three armed men planted two bombs in the Midland Hotel. When one of the devices detonated a fire swept through the ground floor of the hotel, however, fire crews were standing by at the scene

---

[162] McKittrick, Kelters, Feeney and Thornton (1999).

[163] Annual Report for the Fire Authority for Northern Ireland.

and the fire was quickly brought under control. The hotel had been bombed twice before, once in 1972 and again in 1974. It was quite well known to both the Brigade and the ATO:

> A black Austin 1100 was the centre of attention, on the back seat was a bomb we called the Midlander because we'd pulled the first of its type out of the Midland Hotel in Belfast. It contained about 50 lbs of explosive and some tricky anti-handling devices.[164]

> In the mid '70s, the Midland Hotel, where one bomb had exploded, and a certain Divisional Officer at the time decided that if we didn't get this knocked down the building would be lost. So he went in along with myself and someone else, in through the main foyer, and half way through this we quickly discovered there was another device, the timing device and the sticks of gelignite sitting on top of the reception. So we got out. We knocked the fire down and the building was saved and the second device was made safe, so it turned out all right but it could have gone the other way.[165]

Mentoring

By this stage of the Troubles the firefighters of Belfast had considerable experience in the most difficult and hazardous circumstances imaginable. And, in the same way as those who had come through the Second World War helped to guide less-experienced firefighters at the beginning of the Troubles, the more established firefighters of this stage of the conflict guided and looked after the new and less experienced now. One of my early mentors was Billy Little and I wasn't the only wet behind the ears firefighter mentored by him:

> Great Victoria Street, it was a set of shops on fire. So as soon as we got the call round to it jets were put out and we climbed up the staircase with a jet, putting it out as we went. And eventually got up to where these rooms out the back of it were blazing like mad, and it just flashed over on us. Billy and I were up on a jet and I was as naive as anything, when the whole thing flashed across, and Billy jumped over the top of me and knocked me to

---

[164] Styles and Perrin (1975).

[165] Ken Harper.

the ground. It just went over the top of Billy because he was lying on top of me. I had a lot of respect for him after that. And he did the same thing a few years later in the Grove.[166]

On the night of 15 July a large fire badly damaged the Grove Theatre. Firefighters with nine appliances fought for three hours to save as much as possible of the building, even though the roof collapsed at one stage. Despite their efforts the building was badly damaged by fire:

> I was standing beside Billy on a jet at the Grove Cinema fire, we were sent up the stairs with this jet, and we stood right in the vestibule up the stairs where they had the bar and the shop. Well, Billy and I stood in that doorway, when the roof started to collapse on us and I started to run and he grabbed hold of me and he said, 'stand here you'll be alright'. And literally all the debris from the roof finished at our feet and he never moved. That instilled a lot of confidence in me. I had much regard for that man.[167]

On 4 August there were several incendiary attacks in the city linked to a visit of the Queen, planned for the following week. By this stage of the Troubles most of the incendiary devices being used were of the 'cassette' type. They were generally planted in the late afternoon and were designed to detonate within 12 hours. Young women, often between the ages of 17 and 21, were used to plant the devices and they employed a number of methods to do so, sometimes using prams or baby's clothing to hide them. Despite business owners taking the precautions of searching premises and hiring night watchers, the women often managed to succeed in planting these small and easily hidden devices to devastating effect. For example, on 18 August 35 incendiary devices were planted in 25 shops and businesses in the city.

On the night of 5 September there was another wave of incendiary attacks in the city centre. The devices used were once again of the cassette type and they all detonated in shops within the security gates, despite the increased security. On 23 September, a fire started by incendiary devices destroyed the biggest cinema in the city, the ABC. In what was believed to have been a concerted attack, two other cinemas, the nearby New Vic and the Curzon on the Ormeau Road, were also damaged. The devices all

---

[166] Gordon McKee.
[167] Gordon McKee.

detonated at about the same time and firefighters dealing with the fire in the ABC had to extend operations to the New Vic, when smoke was seen billowing from that cinema:

> I was there the night of the incendiary attack on the cinemas in Belfast. The Ritz and the ABC and the Curzon all gutted in the same night. They were really hard fires to deal with because the cinemas were total enclosed buildings with no windows so we had to get inside to deal with the incidents.[168]

Industrial Unrest and the National Strike

Industrial relations in the United Kingdom were poor throughout the 1970s and the fire service was no exception:

> 'The year has again been beset with industrial action', wrote the Chief Inspector of Fire Services in his report for 1975. But by then conflict seemed endemic; relations between firemen and their employers had so deteriorated that it was the absence of unrest that was considered noteworthy.[169]

During that year, on 12 May 1975 the firefighters of Belfast joined their colleagues in the rest of Britain in answering emergency calls only, in a campaign for more pay and shorter hours. There were no sanctions for being on work-to-rule at that time, and as the only work that was done was answering emergency calls, it was regarded as quite good fun at times. The work-to-rule ended on 14 August that year when a £6 a week pay rise was agreed at a meeting in London. However, the calling of a work-to-rule occurred under almost any pretext throughout the rest of the 1970s. Nonetheless, at a time when the national average wage of an adult male worker was £76.60 for a basic 42-hour week. The weekly wage of a firefighter ranged from £52.53 on appointment at the age of 19 to £65.70 for a qualified firefighter with four years training, for a 48-hour week including shift work.[170] In these circumstances tensions between

---

[168] Colin Lammey.

[169] Englander (1992).

[170] Bailey (1992).

management and the workforce continued and frustrations increased as good firefighters left the service for better paid jobs:

> There was a lot of discontent around the service. I came on to White Watch Central and five guys had left, I was one of five recruits, they had maybe lost five or six of their firefighters who had left and joined the prison service or left and joined the police.[171]

As a result of several years of increasing frustration, on 7 November 1977, a Recall Conference of the FBU passed a resolution to go on strike, and within the next few days soldiers started to prepare to take over from firefighters. At 9.00am on 14 November 1977, firefighters and control staff in Belfast, and all over the UK, walked out of their stations and the strike began, although many retained firefighters continued to work normally. Every station in the city was picketed and the army, with a mixture of firefighting specialists and newly trained soldiers, were deployed with green goddesses (military fire engines that had been mothballed for many years).

Throughout the strike the Troubles continued and there were several attempts to kill army firefighters with booby trap explosive devices, some being planted in fire hydrants. The additional hazards to army firefighters in the Northern Ireland context were self-evident, but there could also be additional difficulties for striking firefighters. And, on 22 November:

> We were standing on the picket line with the brazier going at the front door. A black taxi came up the Woodvale Road and stopped. These two guys got out and said: 'There's a row of derelict houses on fire behind the library on the Shankill Road and it's quite close to some pensioners' houses. We would like you to come down and put the fire out.' And we were, 'oh no, no, no, we're on strike'. So he sort of asked us again, nicely, 'would you please come down and try and sort this out for us?' And we were, 'no, no, no, we're on strike'. Then I felt something poking me in the side and I looked down and this big guy had a pistol, a gun. And he just put it in my right side under my ribs and he said, 'I'm not asking you anymore, we want you to come down and put this out'. So, we had no choice. We got on the pump, I drove it down, and there was about four guys volunteered to come, and we put the fire out.[172]

---

[171] Colin Lammey.
[172] Stanley Spray.

In a strange twist, the UVF then made an anonymous financial contribution to the firefighters' strike fund.

> A few weeks later somebody got a letter; it said on it 'Shankill Road UVF, 20 pounds. Sorry for breaking your strike'.[173]

**

I was stationed in fire prevention at the time and picketed Whitla Street station during the strike. At the beginning of the dispute I was surprised at the levels of public support for us. Many people gave money to a strike fund and there were many acts of kindness. One example came from the Midland Hotel opposite the station. One day while I was on picket duty the hotel had to have its beer pumps cleaned and serviced. This was done, and afterwards, in order to ensure that the taste of the beer wasn't unduly affected by the cleaning, a number of pints of beer had to be flushed through the system. We were invited over to help, by drinking the beer that would otherwise be thrown away. I don't remember how many pints my fellow pickets and I got through, but I'm certain that it was many more than was necessary to clear out the bar's system.

**

1978

NINETEEN-SEVENTY-EIGHT WAS DOMINATED by the Maze prison dispute, as the 'blanket protest', which had begun in the previous year continued. The dispute escalated in March when republican inmates, refusing to slop out, began smearing excrement around their cells. This dirty protest involved more than 250 prisoners by the summer.

The Strike is Resolved

As the strike continued, many firefighters were finding things difficult, although the degree of difficulty depended on an individual's personal circumstances.

---

[173] Keneth McLaughlin.

It just went on and on and on and on, nine weeks. There was an awful lot of people just didn't survive the strike, they had to get out of the job. A hell of a lot of good guys left.[174]

I coped with the strike all right. I was still living at home. But I was certainly conscious of many of the married men who were on strike, really, really getting it hard and really very worried about how they were going to provide for their family.[175]

The generosity of various people was greatly appreciated:

So many people coming up and giving us boxes of groceries and stuff.[176]

And from time to time the strike was broken:

They did come up on one occasion and say, 'there's a house on fire down there you had better get down', and the boys said, 'yeah, sure', and they took the machines down and put the fire out and then the army arrived up with the green goddesses.[177]

By 6 January employers had made another offer that would give a ten per cent rise if accepted, more in November and a further rise in 1979, which would bring firefighters wages to the level of skilled manual workers. The union indicated that they could not negotiate a better deal and argued that the introduction of a pay formula was a great victory; however, opinion among firefighters seemed divided. In Northern Ireland mass meetings were arranged in Belfast and Londonderry to vote on the employers' latest offer. I attended a stormy meeting of Belfast firefighters in Transport House, where we voted by four to one to end the strike, 280 in favour and 70 against. There was a five to one vote in favour in Derry. A great deal of bitterness was expressed in the Belfast meeting regarding the Lord Mayor's negotiating position (Councillor Stewart was the Lord Mayor of Belfast and National Joint Council member for the Fire Authority), the few full-time firefighters who didn't strike and the retained firefighters who didn't strike. On 12 January, at a special delegates conference in London, firefighters voted by a majority of three to one to

---

[174] Walter Mason.
[175] Colin Lammy.
[176] Murray Armstrong.
[177] Murray Armstrong.

return to work on Monday 16 January. However, there were angry scenes outside the meeting where firefighters picketed. Fighting broke out before the meeting and a smoke bomb was thrown. Firefighters returned to work on 16 January amid a great deal of bad feeling.

> The strike left a bad taste you know.[178]

> What we went back for really, you'd have took five or six years just to make back what you had lost. You know it really wasn't that good.[179]

And life for those who didn't strike could be made difficult:

> There was a lot of ill feeling because some boys didn't go out on strike. There was a lot of ill feeling with individuals at that time. Some boys just left.[180]

Of course, as it was Northern Ireland, there were risks on returning to work, which didn't exist in the rest of the United Kingdom.

> There was a suspicion of things like hydrants, which could have been booby-trapped for the army.[181]

The strike may have been over but industrial relations did not improve and works-to-rule continued well into the next decade.

The Troubles continued and, at breakfast time on 18 January, two blast incendiaries started a fierce fire in Clifton Street. The devices were at a printer and stationers and an army surplus store next door. There was no warning given but passers-by saw the devices attached to the security grilles on the windows of the two shops and the area was cleared minutes before the first device detonated, starting a small fire. The second device exploded shortly afterwards and started a much bigger fire. Firefighters from Whitla Street moved in, fought the fire and stopped it from spreading, even though the ATO had to deal with an elaborate hoax bomb just yards away. Earlier that morning incendiary devices started a fire in Ross's Auction Rooms

---

[178] Jimmy Armstrong.
[179] Dessie McCullough.
[180] Joe Sloan.
[181] Jimmy Hughes.

in Montgomery Street, off May Street. The fire had a strong hold before firefighters arrived to deal with it, the top two floors of the premises were badly damaged and the roof collapsed into the building. The fire destroyed a number of valuable paintings and antiques.

On 12 February a bomb in two gas cylinders, attached to a can of petrol and hung on the front door of a house in Oldpark Avenue, detonated and started a fire that quickly trapped the woman occupant and her ten-year-old grandson. By the time the Brigade arrived there was little firefighters could do to prevent their deaths, as the fire took hold so quickly. The firefighters in Ardoyne were having a busy night and were just back in the station when the call came in:

> Got back to the station, we were foundered. We reversed into Ardoyne station, I was just going to walk straight into the shower and the bleeps came on and it was, actuated 77, it was Oldpark Avenue, persons reported. Get the gear back on, get the BA sets on, freezing, snowing away, get down onto the top of Oldpark Avenue. There was a granny and a wee boy trapped in the front room. It was terrible. They were dead. It was a blast incendiary, just hung on the door. The worst about it was it blew up the stairs, you know.[182]

By this stage of the Troubles the fire bombers had developed a new type of blast incendiary, by adding an additive to the petrol, usually sugar or a domestic cleaner. This made the petrol more sticky and, following an explosion, it could stick to walls, other structural elements, and indeed people, more easily.

The La Mon House Hotel

On the night of 17 February, approximately 400 people were enjoying a night in the La Mon House Hotel. They included members of two clubs, the Northern Ireland Junior Motorcycling Club and the Irish Collie Club. Collie Club members were at their annual dinner dance in the Peacock Dining Room when, sometime after 9.30pm, a hijacked car pulled up to the building and a blast incendiary was attached to the window security grille with a meat hook. There was a telephone warning but as the call was being taken the

---

[182] Kenneth McLoughlin.

device detonated. The device was attached to four petrol containers wrapped together and the 1lb. explosive charge was encased in a steel tube, which when the device detonated, shattered into steel fragments that were thrown into the air, ruptured the petrol cans and ignited the petrol. A massive fireball, 60ft in diameter and 40ft in height, was blasted into the Peacock Room and the heat wave could be felt 300 yards away. The effect was described as being like a gigantic flamethrower. Twelve people died and more than 30 others were injured, many of them seriously. Most of those killed were seated next to the window where the device was placed, six of them at two tables near the point of explosion. A witness told the inquest about hearing an explosion, seeing a ball of flame and feeling intense heat. He said some people were in flames, the lights went out and there were a lot of fumes, adding: I was on fire and I dropped to the floor and rolled in a ball trying to get air.[183]

Firefighters dealing with the incident could smell the burning flesh. Because of the intense heat some of the bodies shrank and it was at first thought that they were children. It was also very difficult to identify some of the bodies because they were so badly burnt.[184] A woman whose husband died said that they were just finishing their second course in the Peacock Room when there was a ball of flame. She said:

> The flames hit me in the face and arms. I looked round and my husband was on the floor with his coat on fire. He was rolling around trying to put out the flames. I rushed to the door and thought that my husband was behind me. When I realised he was not I was unable to find him because of the intense smoke.[185]

Assistant Chief Fire Officer Sammy Moore described the effects of the fireball within the building:

> When a fire is started by a ball of flame, it starts a very fierce fire as the building construction heats up to ignition point in a short length of time. This drives the oxygen from the atmosphere and quickly you get a tremendous blowtorch effect with flames rolling off the top and fresh air coming in from the bottom. At La Mon a type of firestorm was created with the wind appearing to rise

---

[183] *Belfast Telegraph.*
[184] McKittrick, Kelters, Feeney and Thornton (1999).
[185] *Belfast Telegraph.*

near the building. I have never witnessed this in a single-storey building before.[186]

Some of the injured jumped from windows with their clothes on fire, and curtains were used to cover the bodies of several badly burned people. Members of staff and others tried to contain the fire using hoses and fire extinguishers in a valiant but vain attempt. Others tried to reach the injured despite blinding smoke and intense heat, husbands looked for wives and wives looked for husbands. Strong winds fanned the flames and firefighters tried, at considerable risk to themselves, to find survivors while controlling the raging fire, despite the strong winds spreading it throughout the complex. At 9.51pm, the entire premises were well alight and by 10.15pm the first two bodies had been taken out. The bodies were badly burnt, hardly recognisable as human remains, and the building was totally destroyed.

In common with many people from Belfast the Ulster Hospital at Dundonald holds a special place in my heart. My children were born there, and my mother died there. If you go into the building you can see a memorial window to the 12 people, including three married couples, who died at La Mon. The window depicts a couple embracing and at the top the inscription reads,

> You Say the Name and I See the Place
> La Mon

Two shipyard workers were killed and six others injured in a gas explosion on a ship under construction in the shipyard at about 11.00am on 21 April. The explosion was caused by the build up and ignition of propane gas aboard the vessel. The two workers who were killed were in the double bottom area of the ship. Eight appliances attended this incident, removed the bodies and helped with the injured.[187]

On the night of 27 April, Ardoyne Fire Station received a call to Ardoyne Bus Depot. A police patrol had seen a fire and called the Brigade. On arrival it seemed as if the whole depot was on fire; flames reached up towards the sky and smoke whirled around the firefighters as they ran out jets of water to fight it. The fire destroyed eight double-decker buses and four single deckers. Two other double-deckers were slightly damaged. Various seats of

---

[186] *Belfast Telegraph.*
[187] Annual Report for the Fire Authority for Northern Ireland.

fire were started by incendiary devices, which were planted in the driver's cabs of several of the buses, and a charred petrol can was later found in one of the burnt-out vehicles. On the night of 29 April, a further four buses were destroyed in a separate arson attack. Two 16-year-old boys were later charged in connection with both attacks. By this stage there had been over 300 buses burnt in the city during the Troubles.

**

At the beginning of June I was promoted to the rank of Station Officer on the White Watch in Ardoyne. I was still 24 and even though, in common with most Belfast firefighters, I felt pretty experienced operationally, I was still pretty green as far as watch management was concerned. I was lucky to have Jack Warden as my number two and Hugh McCalmont and George Goodman as my leading firefighters, all three were experienced and steady with many years in the job between them. I also had a watch of firefighters who were always up to something and ready for a bit of fun, but becoming immensely professional when we were turned out to an incident. Over the years that followed there were many personnel changes on the watch, but their operational professionalism never lessened. I felt then, and still do, incredibly privileged to have worked with them all.

**

On the night of 1 June four blast incendiaries were planted in Ormeau Avenue. Two of them were hung with meat hooks onto security grilles over the windows at the Gas Department headquarters. The first device detonated, throwing a ball of flame through the ground floor of the building. The night watchman, who was in the premises at the time, had a lucky escape as he managed to get out of the building before the second device detonated. Firefighters were hampered from fighting the fire as the other two devices had been planted at an electrical wholesaler nearby. The ATO made these devices safe but by the time firefighters were able to make a close attack on the fire, the six-storey, traditionally built office block, measuring approximately 150ft by 150ft, had been destroyed. It took firefighters with eight appliances, including three turntable ladders, about 12 hours to extinguish the fire,[188] but by then all that remained was a pile of smouldering rubble, with the external skeleton of the building standing like something reminiscent of the blitz.

---

[188] Annual Report for the Fire Authority for Northern Ireland.

> We got the gas office, just round from Bankmore Street, and we arrive along and the whole gas office was gone, and that to me was amazing, it was just like Armageddon.[189]

On 9 June, four men were stripping a derelict house prior to demolition in Spruce Street when the roof and part of the first floor collapsed. Two of the men were trapped under the rubble and firefighters had to dig in the debris to free them. One of the trapped men was killed and the other seriously injured.

On the night of 12 June, ten blast incendiaries were planted at the Smithfield Bus Depot. Three armed men drove into the depot, held up staff and planted the devices. There were ten explosions in quick succession and a massive fire was started. In all, 21 buses were destroyed and the depot itself, a range of single and two-storey buildings approximately 350 metres by 200 metres, was badly damaged, with its roof collapsing into the fire. At one stage nearby residents had to be evacuated as firefighters fought the intense blaze. Twenty-five buses, which were parked in the yard of the depot, were driven to safety as firefighters with seven appliances used 14 jets of water, including ground monitors and one from the turntable ladder, to deal with the fire.[190] Nine blast incendiaries were hooked onto the protective window grilles of a string of business premises in Great Victoria Street on 16 July. There was a warning but it wasn't given until after the first three devices had detonated. When the Brigade arrived an area of about 300 metres by 200 metres of the four and five-storey warehouse, which contained the premises, was an inferno. Flames were leaping out of the building and up towards the sky as a column of thick, black smoke rose above the city. Firefighters with six appliances, including two turntable ladders and a hydraulic platform, fought the fire using 20 jets, two ground monitors, two turntable ladder monitors and a hydraulic platform monitor,[191] but there was little they could do other than containing the fire and stopping it from spreading. The blaze destroyed a textile factory, a paper manufacturer and a picture-framing company, and only the external shell of the building was left when the fire was brought under control, some two hours later. A timber merchant, a second-hand car dealers and a surgical supplies shop nearby were also

---

[189] Louis Jones.
[190] Annual Report for the Fire Authority for Northern Ireland.
[191] Annual Report for the Fire Authority for Northern Ireland.

damaged. Three unexploded devices were dealt with by the ATO and the only casualty was a firefighter who was treated for burns.

On 12 October at about 10.30am, four devices exploded on a six-coach train from Dublin to Belfast as it pulled into Belfast's Central Station. A mother of five daughters was killed and four others were injured, three of them seriously. About 100 people were on the train at the time, most of them women on a cheap day excursion. They rushed to get off the train after the first explosion but were still not clear of it when the second device detonated and the train caught fire. At that point there was panic and some people jumped, screaming, onto the track. The second and third devices exploded quickly after that. Wreckage from the train was strewn across the railway line and the train was badly damaged by the explosion and fire that followed. Four coaches were destroyed as firefighters with four appliances used five jets of water to fight the fire.[192]

On the afternoon of 23 October, two bombs exploded in the SPD warehouse on the Springfield Road, starting one of the biggest fires in the city for years. Crews from every station in the city fought the fire, and successfully prevented it from spreading to an 800-gallon oil tank in JP Corry's timber yard nearby. The men who planted the bombs hijacked a post office van to carry the devices to the warehouse, one of them was left beside a consignment of cooking oil and this made the resulting fire burn even more fiercely. At one stage firefighters attempted to fight the fire with high expansion foam but this was thwarted by the high winds, which blew the foam away. The warehouse was badly damaged and thick smoke drifted across the city, as firefighters dealt with the blaze.

Wesley Orr: Death of a Firefighter (II)

At about 8.20am on 16 November a number of armed men pulled up outside the gate of the Ulster Brewery on the Glen Road and held the security guard at gunpoint. The driver then drove several hundred yards to the warehouse, planted a blast incendiary device and returned to the gates to collect his accomplice. The device detonated minutes later. Appliances from Lisburn and Cadogan attended and Assistant Divisional Officer Mick Malone was the officer in charge. On arrival it was found that the device had exploded in a spirit warehouse and a serious fire was developing. Wesley Orr and a

---

[192] Annual Report for the Fire Authority for Northern Ireland.

number of other firefighters had just entered the burning building with jets of water when a second explosion took place; they were caught in the blast. Wesley caught the full force, he suffered multiple injuries and despite an emergency operation in the nearby Musgrave Park Hospital he died several hours later. Five other firefighters were slightly injured. Speaking a couple of years later Mick said:

> I had just left his side and walked 45 feet when a bomb went off. He came out holding his head and said he would be all right, but he died later from his injuries. He was a bit of a character and a very courageous officer, one of the old school.[193]

Just after the detonation the first reinforcing appliance from Central arrived, with Leading Firefighter Harry Welsh in charge, to find injured firefighters staggering from the building and collapsing outside.

> Just as we were driving into the place the bomb went off, so as we were driving up to it they were all staggering out of the place. We couldn't get an ambulance, an army Saracen was the only thing we could get for some reason, I never quite understood that, but we sort of got them all bundled off to the hospital. We were still there about lunchtime and we heard Wesley had died.[194]

I was stationed in Ardoyne at the time and, at 9am, as we were coming on duty, brigade control phoned and told me that there was an incident at Bass Charington on the Glen Road and we were to relieve Lisburn, with me taking over from Wesley. I called the crew together and told them that it looked like we were in for a busy day and to get a cup of tea while the other crew checked the appliances. We would set off to the Glen Road as soon as they were ready. We set off and arrived at the Brewery just behind Central's appliance. As we came over the hill we saw the modern single-storey warehouse building laid out below us, we could see the fire was getting a hold. We came down the hill and it was obvious something was very wrong. Firefighters were staggering out of the building and I saw Wesley being helped out as I jumped out of the appliance. There was little we could do to help him at that time so we started to get a ground monitor and jets on the building from the outside. As it wasn't possible to enter the building at that

---

[193] *Belfast Telegraph*.
[194] Harry Welsh.

time we poured high expansion foam into the warehouse to smother the fire, which partially worked. But we still had to enter the building to extinguish the fire once enough time had passed for the ATO to judge it was safe to do so. [195] I have often wondered what would have happened if I hadn't told the guys to get that cup of tea and we had arrived a couple of minutes earlier.

Wesley had been in the fire service for 35 years and was in charge of Lisburn Station. He had been awarded the British Empire Medal in the New Years Honours List and was well respected in the Brigade:

> When I started in the job Wesley was my mentor if you like. He was a rough diamond, he used to wash the hose amongst various other things, so when you went to a job and you got one of the hoses out Wesley would come over and start yelling at you to put it back and get the hose-reel because he didn't want to wash it. That was his full-time job. And that's the sort of guy he was. He was a real character.[196]

He was buried with full honours from his colleagues, and firefighters from Dublin and Clones were among the many who walked behind the fire appliance carrying his coffin. In November 1985 a stained glass window in his memory was dedicated in the Lisburn Methodist Church. Speaking for his colleagues, Sydney Pollock, the Deputy Chief Fire Officer said:

> Mr Orr was a man who lived and worked in the highest traditions of the fire service. He was a man of courage and was fearlessly leading his men when he was fatally injured. He was dedicated to the fire service and died serving the community.[197]

1979

THE POLITICAL SITUATION in 1979 wasn't promising and a Government proposal for a constitutional conference was rejected in Northern Ireland. The Conservatives won the general election in May and this brought Margaret

---

[195] Annual Report for the Fire Authority for Northern Ireland. Three appliances, three jets, a ground monitor and high expansion foam were used to extinguish the fire. Even so the damage ran into approximately a million pounds.

[196] Harry Welsh.

[197] News Letter.

Thatcher into power. The violence continued and there were many shooting incidents. In October it was announced that Miss Claire Macmahon would be the first female chair of the Fire Authority for Northern Ireland. She had been a member of the authority since its inception and despite breaking the mould she nodded to the culture of the day in her first public statement when she said:

> Perhaps one day we will have women firefighters but the problem is that few would possess the strength needed for the task of holding a heavy hose, for example.[198]

Own Goals

As the Troubles continued, bomb makers and those who planted their devices often died in what became known as own goals, and on 5 January it was my turn to deal with the aftermath of this kind of incident.

I was just finishing off a night shift in Ardoyne at about 7.30am, when we received a call for the pump and pump escape to an actuated explosive device in Northwick Drive, in the Ardoyne area. Apparently two men had been loading blast incendiary devices into a stolen car when one of the devices exploded. We turned out to the incident and arrived at a scene of complete turmoil. I could see a car on fire and what looked like two bodies lying in the road near the car. There were a number of people in the street shouting and screaming.

As I got out of the appliance and onto the footpath a man who seemed to be from the area grabbed me. He took hold of the front of my firefighting tunic and pushed me back against the appliance, shouting: 'Where's the ambulance, get a fucking ambulance, get a fucking ambulance or I'll kill you.' Banging me back against the appliance to emphasise his words. I shouted back: 'Let me go and I'll do what I can.' I then shouted to the driver: 'Order an ambulance on the radio.' I shook myself free and ran over to the two men to see what could be done to help them, as other members of the crew brought the first aid kit and ran off a jet to extinguish the car fire, which was still burning in the street. A quick look told me that the two men, who were both aged 24, were in a bad way and were still slightly smouldering,

---

[198] *Belfast Telegraph.*

however they both seemed to be alive. Members of the watch tried to keep their airways clear until the ambulance arrived and took them to hospital.

My fire report indicated that the source of ignition was: 'Homemade explosives' and the defect, act or omission leading to the fire was: 'Premature explosion of devices being carried by terrorists.' In my fatal fire reports I indicated that it would appear that an explosive device detonated in the boot of the car when the casualty was in close proximity to it. A firearm was found at the scene.

A witness later told the inquest that when he ran to the scene immediately after the explosion he found the two men lying in the street and he tried to comfort one of them who asked for a drink of water. They were taken to the Royal Victoria Hospital but were both dead within two hours. Their injuries were so bad that police had difficulty in identifying them and one of them was wrongly identified at the time.

Argos: A Shameful Episode

Early in the Troubles the owners of premises would occasionally tell the emergency services to take what they wanted from the rubble following explosions, as what remained was so damaged it was likely to be dumped. This 'salvage' at incidents carried on for a while but could not continue indefinitely. In the late '70s a number of soldiers were prosecuted for stealing at incidents and a number of police officers were also accused of this type of theft. On the night of 26 March an explosive device started a fire that badly damaged most of a four-storey multi-occupancy building in Cornmarket. One of the premises concerned was Argos, the catalogue retailer;

> There was a big pall of smoke; I went down to the duty office to listen to the (Brigade) radio. They were saying there was a fire in Argos. Then we got called down, the two pumps went down for relief and we arrived up and this thing was just going like hell. They said the stop message was in, but obviously it's not under control. The first person to meet me was one of the ones who left after it. But as soon as I jumped out of the fire engine this character was there and he just went, 'would you like a watch', and he had about ten wristwatches up his arm. I just said, 'what!' So that's when the alarm bells started to ring with me. I remember at one point going into Argos and it struck me as funny that there were all these boxes lying everywhere, not burnt or anything, just

like the place had been ransacked, which obviously it had been. But I didn't know that at this point. So at that point Raymond Moore came to me and said, 'the fire's gone into this place next door', it was a building society or something, 'will you take your people in there and stop it'. To me that was great because that saved us from having anything more to do with it.[199]

On the first day after there was only three of us on duty and the rest had apparently been arrested. Police came into the station and interviewed all those who had been there, and some of those arrested hadn't even been there at all. And then the fall out from that became quite incredible in terms of the damage that was done and the number of people who left the service. So I suppose I count myself lucky that I wasn't there. I don't think it was a good night for the service at all.[200]

On 3 May, 1979 nine firefighters appeared in court on charges of theft and receiving stolen goods. They were suspended from the Brigade, without pay, and eventually admitted to taking almost £1,500 worth of goods from Argos during the fire. Rings, bracelets, watches, pendants and cigarette lighters had been taken from the strong room after it was forced open. Following the court case, the firefighters concerned were fined and given suspended jail sentences. They all lost their jobs. My watch was on duty that night, although I was on leave, and we had a pump attend the incident. I was always pleased that none of my guys were involved. One of the sad things was that a number of otherwise good firefighters, who had worked hard in terrible circumstances at earlier incidents, got caught up in this collective madness.

The MV Inio: Walter Wilson and the Magnificent Seven

On 10 April, the Helsinki-registered 2,500 tonne freighter, the MV *Inio*, got into difficulties off the coast of Northern Ireland when a fire was reported below her decks. The skipper called for firefighting assistance and crews, initially from Whitla, attended to deal with the incident.

---

[199] Harry Welsh.
[200] Colin Lammy.

We received the call at just around 1pm. Myself and Grant Ashcroft and Dessie McCullough I remember were the breathing apparatus wearers. So we arrived down at the tug and when we got there we found that there was a fire somewhere off the coast of Northern Ireland in the engine room of this ship called the MV *Inio*. We took a lot of equipment with us and we got onto the tug with our breathing apparatus sets and off we went. There were eight of us; Walter Wilson was the Station Officer and seven of us with him. We got out of Belfast Lough and then we all started to settle down because it was a good hour and a half for the tug to get to the ship. It was quite a swell that day but it was a good bright day, and the ship was probably about 30 or 40 feet and going up and down. So we realised very quickly that if you were going to get on the ship, then you had to jump whenever the tug came up with the waves to the level of the ship's deck, because if you didn't you would fall right back down about 30 or 40 feet. So whenever the tug came up I jumped and there was a cargo net on the side of the *Inio* and I remember aiming for that. I'd my BA set on and John Hunt put his hand out to grab me, as I lost my balance, and he got me by the back of the cylinder and pulled me onto the ship. So we fought the fire and there were only eight of us, and of course nobody else was coming. After that we were called Walter Wilson and the magnificent seven.[201]

Dessie McCullough also remembers this incident:

That was a good wee fire too. It was in the galley. I was just going up to the galley when there was this unmerciful explosion, they had two big gas cylinders to feed the cookers and they both exploded at the same time, both of them. They were inside this big steel cupboard thing; the whole thing just lifted right off the deck into the water. But we put the fire out anyway. We saved the whole cargo and we actually got salvage money, which was the first time the Fire Brigade ever got that.[202]

**

---

[201] Louis Jones.
[202] Dessie McCullough.

The bombing and burning continued and, on 26 April, three men entered the tax office in Ormeau Avenue. One of them held a security guard at gunpoint in the entrance lobby while the other two planted two blast incendiaries on an upper floor. Over 100 people managed to escape from the five-storey traditionally built building before the devices detonated within minutes of each other and started a fierce fire, which quickly spread throughout the premises:

> The motor tax office in Ormeau Avenue is just about where the current Bankmore Street Fire Station sits. Several incendiary devices had been planted in it. The people had been evacuated when we got the call. I was on the TL that day and it was the old Metz TL. If you were the third man on the Metz you sat outside on the appliance and had to hold on as it went round the corner for fear of getting thrown off. No seat belts, no nothing. And whilst you were holding on, you also had to get dressed in your fire kit and get your TL belt on. The bombs went off in the building, five or six in fairly quick succession and there were one pound and five pound notes blown out of one of the windows after one of the bombs had gone off. It was like confetti coming down around us. By the time we got the TL set in there were flames coming out of every window. I was up this TL, of course no breathing apparatus up there in those days, shot up by the operator as high as he could get me, he then dropped me down in towards the window so these flames were rolling out over the top of my head and then he had to stop, run round to the back of the fire appliance hit the power take off and allow the water in and up the ladder. [203]

> This place was going like the clappers, jets all round the place as usual, three or four pumps out the front. Nowadays you would have virtually every resource in the Brigade there you know, and we had four or five jets out and a couple of people were inside searching, not searching for people because we knew it had been emptied, searching for other avenues of where we could get a jet in and contain this thing. [204]

---

[203] Colin Lammy.

[204] Murray Armstrong.

A New and Improved Explosive Mix

By this stage of the Troubles the IRA had managed to refine the mix for fertiliser based explosives, in order to make them more reliable and efficient. The technique, which involved grinding down the original granular mix to powder form, and making some chemical changes to it, also made the mixture more powerful. Therefore the new devices were more powerful, more reliable and less dangerous to those who were using them, with less of the device being left for forensic examination. The new mix was used on the night of 11 May when a 500 lb. van bomb detonated beside the tax offices in James Street South. The device caused major damage to every building in the narrow street and the blast was heard almost five miles away. A telephone warning was given to the Samaritans and the area was still being cleared when the bomb exploded. No one was injured but six people needed treatment for shock.

The night of 8 August saw the start of the, by now usual, street violence commemorating the anniversary of internment the following day. There was stoning, petrol bombing and a number vehicles were hijacked and set alight on the Falls Road. On 17 October there was chaos in the city as loyalists protested in a demand for the release of six men being questioned by the police. There was hijacking and burning, shooting, three malicious fires and several bomb hoaxes. That night I attended several of the incidents with appliances from Ardoyne station, including two of the malicious fires. The first was at Cairnmartin Secondary School where the science laboratory was badly damaged and we used four breathing apparatus sets and a jet to extinguish the fire. The second was in offices used by the Health and Social Services Department on the Shankill Road. A skylight had been broken and the contents of the offices were vandalised before our arrival. There were two seats of fire in separate offices and the premises were badly damaged before we could extinguish them.

1980

THE TROUBLES RUMBLED on throughout 1980. For the Fire Authority industrial relations remained a prominent issue and a second national strike was threatened over the refusal of the national employers to honour the pay agreement, which ended the strike of 1977/78. Eventually it was agreed that the pay rise indicated by the formula, 18.8 per cent, would be paid in full but over two stages during the year. To be fair to the authority the dispute was not of their making and their view, indicated in their Annual Report, was that the agreements, 'revocation without notice or debate was a major disaster in the field of industrial relations'.[205] Additional local issues in dispute were the transfer of personnel from Ardoyne to Springfield stations and a complaint that the union had not been consulted regarding the introduction of new breathing apparatus. On 18 August firefighters in the city started a work-to-rule over staffing levels and the authority's decision to remove the station based watch room attendant, which the union said would mean a reduction of forty-eight jobs. Firefighters also continued to refuse to move to the new Westland Station, which was being used as a training centre, until the dispute was resolved.

Bombs on Trains

Over the years of the Troubles the railway system in Northern Ireland was frequently attacked, and on 17 January one of the worst of this type of incident occurred at the railway station in Dunmurry. At about 5pm, the Ballymena to Belfast train was approaching Finaghy on the outskirts of the city when an incendiary device, of less than five lbs of explosives attached to a can of petrol, detonated and sent a fireball through one of the carriages, killing three people. All three of the dead were in the rear coach of the two-coach train and their remains were so badly burnt it took some time to discover whether they were male or female. The train conductor said:

> There were flames everywhere and thick, black smoke billowing through the door. The flames were even coming out of the heater grilles. It was a living nightmare and I knew someone was seriously hurt. Firefighters arrived almost immediately and they worked like Trojans in spite of the flames and heat.[206]

When firefighters from Cadogan arrived the carriage was a ball of flame and they used a jet of water to fight the fire, working through the carriages of the burnt out train to finally extinguish it. One firefighter, James McAllister, chipped a bone in his ankle but stayed to fight the fire for another two hours. Two more people were badly hurt and a further five were treated in hospital for their injuries. One of the dead was a member of the IRA and he had been

---

[205] Annual Report for the Fire Authority for Northern Ireland.
[206] News Letter.

carrying the device when it detonated. His accomplice was badly burnt with scarring to 20 per cent of his body. The other two fatalities were a 17-year-old schoolboy and a 35-year-old man.[207]

Just before midnight on 31 February a gang of about ten armed and hooded men entered the bus depot at Ardoyne; they went from bus to bus planting incendiary devices before running away. The devices detonated quickly after they were planted, starting several fires, which spread through many of the buses. In all 20 buses were destroyed and another 11 were badly damaged. I was the officer-in-charge of Ardoyne station that night and it looked as if the entire depot was on fire as the flames rose into the night sky. The fires were quickly extinguished, although we had to stand by until all the remaining devices had been defused. In all 1,240 square metres of the terminus were involved in the fire. William McClay remembers that, Ardoyne Bus Depot was 'forever being targeted… hitting buses, putting bombs on them and setting fire to them. The buses went up in flames in no time'.[208]

A Difficult House Fire

On 27 April, I was on duty in Ardoyne when we got a call to a house fire on the Glenalena Road. When we arrived, three minutes after receiving the call, the house was well alight and I was told that although two of the sons had jumped from a window there were still family members in the house. The watch quickly got to work and we soon had two jets on the fire, one going inside the house from the front door and one from a ladder that was pitched to a first floor window. The fire had such a strong hold that there were flames coming from all of the windows and I wasn't confident that we could save anybody who was still in there. We brought the fire under control and once we did we found the bodies of the owner's wife, his son and his granddaughter in the remains of the house. The wife and granddaughter were together in an upstairs bedroom. As I was gathering the necessary information for the fire report, the owner approached me and said that he had tried to put the fire out with his jacket as his wife went upstairs to get their granddaughter, but it was no use. He then walked quickly away to the rear of the property. Billy Easton, who

---

[207] McKittrick, Kelters, Feeney and Thornton (1999).

[208] William McClay.

was standing near to me, said that there was something wrong about this fire. He confirmed my own suspicions about the demeanour of the owner of the house, whose behaviour and bearing did not seem to fit someone who had just lost three family members, so I informed the police of our concerns. It later turned out that the man had started the fire himself, after an argument with his wife about family members smoking in the house. He had been drinking that night from about nine to midnight. When he returned home one of his sons made a cup of tea and went to bed, from where he heard his mother and father arguing downstairs. He heard his father shout: 'I'll burn you out'.[209] Then he heard his father crumple paper and sprinkle lighter fluid around the house. Smoke filled his bedroom and when he opened the door he saw flames coming up the stairs. He woke one of his brothers but couldn't wake the other one. The two brothers jumped from their bedroom window and tried to go back in to rescue the others but couldn't get into the house due to the heat of the fire. As the two brothers jumped from the burning building their father threw stones at them and shouted: 'Burn you bastards burn'.[210] The man's wife had gone upstairs to try to rescue their granddaughter but she had been overcome by the smoke and fire. The man, who had a history of alcohol abuse, was later sentenced to life in prison for manslaughter. In my witness statement to the police I stated that:

> When we arrived the house was well alight. I was informed by a person that there were people trapped on the first floor. I ordered two of my men to enter the house wearing breathing apparatus and to make their way to the first floor. I ordered water supply to be put into operation. After a period of time I entered the building and went upstairs with Sub Officer Easton and found two bodies in the back bedroom on the first floor. We then went into the front bedroom and found another body. I could only identify one of the bodies and that was a female in the back bedroom. The other two bodies were so badly burnt I was unable to distinguish what sex they were. We stayed there until the fire was extinguished and the bodies were removed. In my opinion the fire started in the front downstairs room.

---

[209] *Belfast Telegraph.*
[210] *Belfast Telegraph.*

Buck Alec's Lion

In November it was announced that a book was going to be written about the 'street fighters' of Belfast. The tales told of Belfast's past, particularly in working class areas, were full of stories of the city's street fighters, who would take on all comers. The fights took place on the street while local people cheered for their favourite, even though some of the stories may have been somewhat embellished. Buck Alec was said to be the champion street fighter of Belfast, he was also a formidable opponent inside the ring, with over 300 professional boxing bouts to his credit. Alec, whose father was a lion tamer, once had three lions himself: Roger, Sheila and Joey. He would wrestle them before boxing matches in the old Chapel Fields in the York Street area and he was once a familiar sight walking his lion as if it was a dog. He was mauled by his lions twice and once nearly lost an arm. By 1980, at the age of 78, he was living a quiet life but still proudly boasted that he was never beaten in a street fight.[211] The Troubles, of course, put an end to this kind of 'hard man' in the city.

The rumour of a book rekindled a half forgotten memory. As a young firefighter I remembered an incident in Tigers Bay, probably around 1972/73. It was a report of a fire in a two-up two-down terraced house. When we arrived there was no obvious fire in the house, so the officer-in-charge directed me and another firefighter to go around the back with a short extension ladder, to climb over the back wall and see if we could see anything from there. Having pitched the ladder to the back wall of the yard, I footed it and my partner climbed it, being careful because of the broken glass set into the top of the wall for security purposes. He reached the top of the wall and looked over. Almost immediately he came back down the ladder, so quickly he nearly knocked me over. 'There's a lion, there's a lion!' 'There can't be.' I said. 'There is and if you don't believe me go up there and have a look yourself.'

So up I went to the top of the wall and carefully looked over. There, on the other side, standing upright with its front paws two thirds of the way up the wall was indeed a lion. It wasn't very young and some of its teeth were missing but it still looked pretty scary to me. So we decided to go back round to the front of the houses and report back. When we did the Station Officer didn't believe us, and the actions we had taken were then repeated, the only difference was that it was the Station Officer who went up and looked over the wall. 'It's only Buck Alec's lion,' he said and we went back round to the front, confirmed that there was no fire by looking through the windows, and left the lion to its own devices.

1981

IN 1981 THE prison dispute developed into a hunger strike during which ten republican prisoners died. As a result there was an upsurge of violence and many more deaths on the streets. Industrial relations continued to be difficult in the Brigade and even though the new station at Springfield Road was officially opened on 28 March, it was not possible to open the one at Westland, due to on-going disagreements over staffing levels. During the year the Fair Employment Agency issued its findings on a complaint of discrimination on religious grounds, made by a firefighter who had failed to gain promotion. The agency found in favour of the Authority but indicated that it would carry out a general equal opportunities enquiry within the Brigade.

On 26 January masked gunmen entered a wholesale jeweller's in Kent Street and ordered members of staff to lie on the floor as they planted a device on the counter. Another man entered and planted a second device, with a petrol can attached, in the showroom. The first device detonated at about 10.00am starting a fire and the second some ten minutes later. Two firefighters, who had begun to fight the fire, were slightly injured during the second explosion and firefighting was then held back due to the fear of further devices. The fire caught hold and the premises were badly damaged before it was extinguished. At 9.00pm that night another device was left at the front door of the same premises, when it detonated it started a fire that virtually destroyed the building. The burning and bombing continued.

From Ardoyne to Springfield

Even though the move to the two new stations that were to replace Ardoyne remained a matter of dispute between the Fire Authority and the FBU, we did move from Ardoyne to the new purpose built station on the Springfield Road. The station officially opened on 28 March with the actual move having taken place on 11 November 1980, sweetened by a relocation allowance that would give us a few pounds extra per week for five years. The leaflet produced by the Fire Authority to mark the occasion indicated that the station would be officially opened by Miss Clare Macmahon, BSc, MBiol, Chair of the authority; on the day the duty watch would parade on the forecourt and the 'Chairman' would ring the fire bell to declare the station open (the bell used was the bell that had hung at the original Fire Brigade Headquarters at Great Victoria Square and was used for summoning part-time firemen in Belfast up to 1871).

My watch happened to be on duty on that day, so we would have the honour of being present when the station was officially opened. Sam McIntyre, who was in charge of the Training School at the time, visited the watch on the previous shift to make sure that we would parade to the required standard. We were told that the Chair liked to be called Madam Chairman and that was the phrase we would all use to address her. On the day I lined the watch up, although we paraded in the engine room as opposed to the forecourt, and reminded them that I would bring Miss Macmahon down the line and introduce her to them one by one. I emphasised that on being introduced they would reply: good morning Madam Chairman. This was duly done, although I am pretty sure that George Goodman referred to the Chair as darling (I think he got away with it). The formalities went well and we all trooped up to the recreation room for light refreshments. Clare Macmahon came across as a very friendly, engaged and interested Chair and, as I looked around the room, I could see that she and George were deep in conversation. I watched the guys eating as many of the cream cakes as possible, there seemed to be some sort of competition going on to see who could eat the most, and the chocolate éclairs were squeezed with the resulting line of cream being licked off. Springfield was a great new station with all mod cons but as I stood there I couldn't help reflecting on my time at Ardoyne, a special station by any measure, busy and right in the middle of things.

> Everybody wanted to join Ardoyne because Ardoyne was busy.[212]

> There were some nights in Ardoyne when the riots were going on, we just closed the gates and sat there watching them.[213]

But most of all I remembered the characters, and while there were many of them, two in particular came to mind.

---

[212] Wesley Currie.

[213] Walter Mason.

Harry the Dog was a firefighter on Blue Watch Ardoyne and he was a well-respected, effective firefighter at an incident, but also a bit of a 'rough diamond'. The most well known story about Harry was the one about the beer kegs:

> Blue Watch were pretty renowned for their drinking ability. They knocked the inside bricks out of a big cupboard in the locker room and they were able to fit two kegs of beer and a cooler in, and it ran day and night. Spike Milligan; he was in charge of the 'wet canteen'. So on the first day of every set he would drive his Mini into the yard and drive it into the tangee shed and his wee trailer on the back. The two kegs of beer were covered in a chequered blanket. Billy Skelly said to Harry the Dog and myself, 'come on round and give us a hand with these kegs of beer'. So we walked round into the tangee shed and Spike had taken the tow bar off the back of the car and set it on the floor. We took the blanket off and Harry straddled the tow bar with his legs and he lifted the keg of beer closest to him. Now there is some weight in a keg of beer, but Harry lifted it right up onto his belly and the weight of the other keg of beer made the tow bar rise and the tow bar hit him right between the legs, right in the plums. Bam! Such a biff, so Harry was standing on his tip toes with his eyes bulging out of their sockets, holding this keg of beer. We were laughing our heads off and Skelly says, 'Harry set the barrel down'. So Harry set the barrel of beer down and instead of stepping over the tow bar he put his two hands on the end of the trailer, he pushed it down, and when he pushed it down the other keg rolled down the trailer and pinned his fingers to the trailer. He let go and the tow bar came up again and hit him again. So he got a double dose of the tow bar in the plums. It was fantastic. It was such funny thing. Poor old Harry![214]

Pat Delaney owned the newsagent's shop opposite the station at Ardoyne and he was a good friend to the firefighters who were stationed there. He was a big guy and was affectionately known as Fat Pat. I can still see him at card games on the station affecting a green visor and smoking a big cigar.

---

[214] Stanley Spray.

Pat was great because of the way you were able to run across there and get your wages cashed, like going to the bank it was terrific. He was a good friend to everybody on the station; he was always in and out of the station at steak dinners and whatever else.[215]

And we were able to repay him for all his favours and support to a certain extent when he became ill, and after his death:

He had taken ill and the ambulance had arrived. But they couldn't get him down the stairs because of his size; they just couldn't manoeuvre him down. So the Fire Brigade got the hydraulic platform across, got him onto it, down into the ambulance and away. Now he was several days in the hospital and he took another turn and he died and we went across to the funeral.[216]

At the funeral the coffin wouldn't go down through the opening to be cremated. It was too wide. By this time the mourners had been led out and the fire service personnel who were there went across to do what they could to help the undertakers and crematorium staff.

We thought we could turn him over on his side. So we took a quick measurement and to turn him on his side, it was nearly the same, it wouldn't go through. Jack Warden said; 'We could take those mouldings off and the thing should go down'. So they went and got the hearth kit and brought it back in, we took all the mouldings off the coffin and it nearly went down, but not quite. If we took the lid off and turned him on his side, because the lid itself was a few inches, it would be thinner and we would get him down. So we took the lid off, turned him over on his side, we were holding Pat into the coffin, and we brought him down. All we were worried about was maintaining some decorum, for want of a better word, because this was somebody that we knew and respected. So we kept him in and we got him down with no one the wiser.[217]

** **

---

[215] Murray Armstrong.

[216] Murray Armstrong.

[217] Murray Armstrong.

The 1981 Hunger Strike

On 5 May Bobby Sands, who had been elected as Member of Parliament in April, died on the 66th day of a hunger strike, which was instigated as part of a campaign to secure political category status for prisoners in the H-Blocks of the Maze Prison. Word of the death quickly spread through republican areas, widespread rioting broke out and the two communities in Northern Ireland became more polarised than ever. There was sustained petrol bombing, buses and other vehicles were burnt and used as barricades and there was a lot of shooting. Businesses were closed, the army fired hundreds of rubber bullets, and several buildings were set on fire. The Brigade was exceptionally busy. Ardoyne seemed to be worst hit: burnt out cars, lorries and vans littered every street in the area and a JCB digger and a dumper truck had been overturned and set on fire in Flax Street. The shooting, rioting, stoning, petrol bombing and acid bombing continued over the next three nights:

> At the back of Springfield, you used to be able to walk out the back like a veranda, and there used to be rails and you used to look over the whole of West Belfast. Another firefighter and I were both on the railings looking out. Then we just heard the bin lids starting to hammer, hammer, and within 30 seconds the alarms went off and the whole place went mad, the whole of Belfast went mad. The first call we got was down Roden Street, and what they did is they burnt a lorry and they pulled it into the middle of Roden Street. We were sent down and we got a couple of jets out. The next thing a guy came round the corner of the road with a Thompson machine gun, we were that close I could see it was a gun with a barrel with holes in it, and he let a rattle off. The noise of it, and we just dove right onto the floor. That set the scene for the whole night. We just went from one call to another.[218]

> We were standing on the balcony; there was just an awful atmosphere. And then it started up the road with petrol bombs and the carry on with the army. We were in the middle of it, cars on fire, lorries on fire and petrol bombs bouncing off the side of the fire engine, it was crazy, absolutely crazy stuff.[219]

---

[218] Wesley Currie.
[219] Kenneth McLaughlin.

We were in the New Lodge and the whole place was bloody going. It was incredible and the atmosphere was just electric. There were bullets flying everywhere.[220]

Yet even in these, the most difficult of times, firefighters found things to smile about:

The last call of the night was in JP Corry's, we had just got to bed about half six or seven, put our heads down for 15 minutes and the next thing the call came up, JP Corry's is on fire. We arrived up there, Jack Warden was the Station Officer, Big Jack, and what they had done was they had broken a breeze block wall that backed on to the Turf Lodge and they had put a hole in it and they were lobbing petrol bombs into all the timber racks. I remember when we arrived, we were just trying to get a jet on all the timber, and I remember Corry running about, and he actually had a gun, so he had, and the army were trying to stop him shooting the guys with the petrol bombs. (I heard a great story about that. The story I heard was Jack was at the pump escape putting a message back and somebody came running across to him and said, 'Corry's got a gun! Corry's got a gun!' And Jack, thinking that as your name was Currie, you were the Corry they were talking about, said: 'Tell him to put it away and get back to fighting that fire.') That's right I didn't know about that until the next night and Jim was telling the story and Jack Warden was laughing about it and he said, 'I thought it was you who had the gun.'[221]

Hijacking Fire Appliances

Fire appliances were hijacked from time to time during the Troubles:

The people who took the fire appliance off us were all holding guns and various weapons. This was in the Clonard Street area and they took the fire appliance off us and it was later used as a barricade. We all walked up the street and took our firefighting

---

[220] Louis Jones.
[221] Wesley Currie.

kit back to the station, but no fire appliance. It was burnt out somewhere down the Lower Falls.[222]

I remember coming down the Shankill to the Stadium Cinema and a crowd came around us and took the appliance off us. But they couldn't get it started and they called us back, 'here you can have it back'.[223]

We got the fire engine taken off us at one of the streets off the New Lodge. A guy appeared up with a gun and ordered us all out. They didn't seem to have any interest in us; they just wanted to use the fire engine as a roadblock.[224]

In the early hours of 9 May, five masked youths boarded a fire engine at Edlingham Street, as it was attending a fire in a petrol bombed furniture and carpet shop during the rioting in the New Lodge. The driver was forced to drive the appliance a short distance before he was released, it was driven across the road and petrol bombs were thrown at it in an attempt to turn it into a burning street barricade. The fire didn't really take hold and the army later recovered the appliance. Later that day the FBU imposed a ban on Fire Brigade attendances to troubled parts of the city, unless life was at risk. We were somewhat reassured when the Chair of Provisional Sinn Féin in Belfast condemned all attacks on essential services, including the Fire Service. He went on to say: 'We appeal to young people not to interfere with essential services, and the Fire Service is an essential service'.[225] We continued to attend every incident.

On 12 May another hunger striker died and we were, once again, very busy. There was violence on the streets virtually every night over a six-month period, as the hunger strikers died. My watch was on duty on the days of the 12th and 13th and the nights of the 14th and 15th. The early hours of the 16th were particularly busy, and a large riotous crowd, who threw stones and petrol bombs at us, prevented us from getting to one of the calls, a number of vehicles burning on the Crumlin Road at Flax Street. On the 21 May another two hunger strikers died as the street violence in all its forms continued.

---

[222] Ken Harper.

[223] Ken Spence.

[224] Harry Welsh.

[225] *Belfast Telegraph*.

Early on 1 June, I attended a house fire in Harrybrook Street, with appliances from Springfield. The fire had started in a chair in the living room on the ground floor, and following the inquest I recorded the probable cause as a 'carelessly discarded cigarette'. The fire was going well when we arrived and the house was severely damaged. A 69-year-old man was killed and his son received burns to the face and upper body before he was rescued by firefighters, who entered the building wearing breathing apparatus. A woman escaped through a back window of the house with help from the Brigade. Three firefighters, Billy Falls, Ken Shields and Marty Magill were injured while fighting the fire and carrying out the rescues. In my witness statement I said:

> We arrived at Harrybrook Street at 2.24am and found the ground floor of the house well alight. I ordered two of my men to enter and search the house wearing breathing apparatus. I ordered water supplies to be put into operation. I made my way to the rear of the house where my men were assisting a female person from the house. This person told me that her husband was still in the house and the last time she had seen him he was in the front room first floor. I entered the building with Leading Fireman George Goodman and saw a male person on the floor in the front room first floor. This person was removed by Fire Brigade personnel and taken away by an ambulance.

This formal account really doesn't reflect the urgency and physicality applied in this type of incident by the fire service, but, hopefully, you get the idea.

PCBs

PCBs (the chemical compound polychlorinated biphenyl) had been used as insulators or coolant fluids in electrical apparatus since the late 1920s, but their use was banned or severely restricted in most countries during the 1970s and 1980s, due to the harmful effects on human fertility, the immune system and the increased incidence of certain cancers in workers who had been exposed to it. Concern was raised when it was discovered that one of the incidents the Brigade dealt with on the 5 August, a car bomb at Donegall Square Mews, involved an electrical transformer, which released PCB following the explosion. As the liquid was leaking, firefighters were tackling a fire in a first floor office at Claredon House, also caused by the car bomb.

Some months after the event an MP who wanted the use of PCBs to be controlled, raised the incident in the House of Commons, and following this, new procedures were agreed between the FBU and the Fire Authority to maintain records of the exposure of firefighters to hazardous substances at chemical incidents.

> Questions were asked in the House of Commons about incidents with PCBs and one of them was an explosion just at the back of the City Hall where PCBs were involved. It was at that stage that we as a union insisted that the chemical log of firefighters be set up and that it would be kept for any chemical incident that any firefighter was involved in. So if someone for example developed asbestosis or those sorts of diseases as a result, then it could all be traced back to when they were exposed.[226]

On 20 August the tenth and last hunger striker to die in the 1981 hunger strike passed away, the violence continued. On 3 October the hunger strike was called off. However, the bombing and burning continued to the end of the year.

1982

AFTER THE TURBULENCE of the previous year the Government pushed for a fresh political initiative, and in April a new Northern Ireland Assembly was proposed. In the elections to the Assembly in October Sinn Féin took ten per cent of the vote on their first serious outing at the polls. However, neither nationalist party took their assembly seats. In the Brigade industrial relations remained difficult, and even though the new fire station at Westland Road had been completed for two years, the on-going industrial dispute over staffing levels meant that it only opened partially on 24 May, housing one appliance. On 27 April following the retirement of Chief Fire Officer George Morrison, after more than 40 years service, it was announced that Clive Halliday, the Deputy Chief of Staffordshire, would be the new Chief Fire Officer for Northern Ireland. Shortly after he took up his post on 1 May, he addressed a seemingly small issue, but one that was important to firefighters. Since the inception of the new Fire Brigade in Northern Ireland the title of the Brigade had been the Fire Authority for Northern Ireland (FANI) and it seemed more appropriate to us when he renamed the Brigade the Northern Ireland Fire

---

[226] Jim Hughes.

Brigade (NIFB), particularly as the title was on all of our cap badges. He also directed that a wide-ranging review of fire cover be instigated,[227] a review that would have far reaching implications.

A New Type of Incendiary Device

1982 saw an innovation in the type of incendiary device being used. The latest devices used a different type of flammable material, which would cause a small explosion and could be lethal when detonated. If someone was foolish or unlucky enough to handle one they could have a hand blown off and/or be badly burned. The materials that were now being used would stick to the skin and produce a violent reaction if they were put into water. The device was small enough to be contained in a cassette-tape box, or something similar, and would be planted in or next to flammable material in the target premises.

> It used to be that if you found one you flushed it down the nearest toilet. I remember it well, we didn't know they had changed the type of device and we went up and found one, I threw it over to Charlie Black and he went into the toilet put it in, and the toilet exploded![228]

A new type of blast incendiary had also been developed. These were disguised as shopping items, with for example a packet of washing powder containing the explosive, and a liquid soap container filled with flammable liquid placed beside it. They were often placed in a shopping bag and left in the shop. Taken together these two new types of weapon made the destruction of premises through fire much easier. Overnight on 9 April, and into the 10 April, there were a number of incendiary device attacks on furniture shops in the city. Fires were started by devices that activated in McCune's on the Shore Road and Kane's in Botanic Avenue. They were extinguished by the Brigade but only after both sets of premises were damaged. Six other devices were found in two other furniture shops in the city centre and in all a total of 19 devices were found and dealt with in different premises over the 24-hour period. That night an announcement in the Arts Theatre, that the theatre should be evacuated due to a bomb scare, made the audience laugh. The

---

[227] Annual Report for the Fire Authority for Northern Ireland.
[228] Jimmy Armstrong.

show going on at the time was 'Up Ulster' and people thought it was part of the act. The administrator of the theatre had to go on stage and make the announcement for a second time before it was taken seriously.

On the afternoon of 25 June, 25 people, mainly nurses, were injured when a 200 lb. car bomb detonated in Brunswick Street. A fierce fire started in the Brolac Paint Shop as a result of the explosion and considerable damage was caused before firefighters could extinguish it. There was a warning that the bomb had been planted but the police were still evacuating the area when the device detonated. A number of police officers were injured and a man walking past was badly injured by flying glass and debris. A number of nurses, who were working the night shift and sleeping in a nurses' home in the street had to be evacuated, and several passers-by were also injured.

> Well we pulled up, Geordie Roundtree was my LF, the place next door was the nurses' sleeping quarters. A paint store and it spread, we rescued 32 that day, getting all the nurses out. We must have had about five jets on it, we had a make up, we made pumps seven, and the TL was there too.[229]

Even as the Troubles continued there was, from time to time, a demonstration of people's basic decency and courage, which could only impress the firefighters who witnessed it. One such example happened on the night of 3 September when there was a fire in a house in Old Lagan Street. The family in the house escaped but a 65-year-old woman neighbour heard a baby crying from the house. She went into the building, which was still on fire and smoke logged, to rescue the child:

> I went over to the house and I heard the baby crying. I went to go up the stairs but I knew the sound wasn't coming from there. It was dark and the smoke was belching black. There were flames at the back of the house. I went in and saw the baby lying on the floor downstairs and I just lifted him. The smoke was bad.[230]

When firefighters arrived they quickly extinguished the fire, but it was the woman who undoubtedly saved the child's life.

On 27 September, a soldier was killed as he unlocked and opened one of the security gates dividing Protestant and Catholic areas. The corporal

---

[229] Jimmy Armstrong.
[230] *Belfast Telegraph.*

was opening the gates at the junction of the Springfield Road and West Circular Road when a planted device detonated. The army regularly opened and closed these gates and children from a nearby estate often played on them. A witness at the scene said:

> I saw him go up in the explosion. It was horrific. The soldier opening the gate was flung up into the air like a rag doll then there was a lot of black smoke and soldiers running everywhere.[231]

Springfield Fire Station was just about opposite the gates and the detonation was witnessed by the firefighters on duty:

> This always stuck in my head. Every morning, you are in the watch room, the soldiers came down with a Labrador dog. The soldier got out sent the dog over to the barriers; the dog went over to the barriers, had a sniff, said this is kosher; the soldier came over and opened the gates. This morning we were sitting there in the watch room, the soldiers came down tied the dog to the Land Rover, a soldier went over himself, opened the gate, Kaboom! Away he went. The wee soldier was dead. So Davy Page went over to help but the fact that the soldier's flak jacket was in our front garden indicated that he was dead.[232]

1983

THE TROUBLES CONTINUED throughout 1983 but at a slightly reduced level in comparison with previous years.

On 3 February a four-year-old boy was swept away by water in an open sewer. The four ft wide brick tunnel carried fast flowing sewage and it lay 12 feet below ground. However, it had been opened for repairs some time earlier, and the hoarding that prevented access had been blown down in high winds the previous weekend. Firefighters attended the scene and assisted police divers in an unsuccessful attempt to rescue the boy; his body was recovered the following day.

On 21 February, flames were seen coming from the upper floors of the Miss Diana Boutique in Royal Avenue. The fire spread to the upper floor

---

[231] McKittrick, Kelters, Feeney and Thornton (1999).
[232] Kenneth McLaughlin.

of Harry Corrie's fabric shop next door as a pall of smoke and fire rose from the roof of the building. Firefighters with 12 appliances, including two turntable ladders, whose operators were at times hidden in the smoke as sparks fell on their colleagues below, took an hour and a half to bring the fire under control. A firefighter was slightly injured when a ceiling fell on him and the upper floors and roof of the building were badly damaged.

On the night of 29 May several blast incendiaries attached to window security grilles in Great Victoria Street detonated and started a number of fierce fires. Three buildings were virtually destroyed and a firefighter was slightly injured. The fact that there were a number of devices planted, and some of them had not detonated, made firefighting operations very difficult. Firefighters dealing with the incident located a number of devices, which had been planted at the various premises.[233]

The Westminster Election was on 9 June and to mark the occasion an oil tanker primed with four or five lbs of explosives erupted in a ball of flames at the police station in Suffolk. The fire rose for several hundred feet into the air as firefighters fought to bring the blaze, which burned for an hour and a half, under control. The driver of the tanker had been forced at gunpoint to bring the vehicle to the police station and as he approached he shouted a warning, which probably prevented any injury at the incident. Late in the evening of 13 June, blast incendiaries badly damaged two women's clothing shops inside the security barriers in the city centre, Evans and Graftons, both in Castle Lane. High winds drove the fire and it was only a quick firefighting attack that prevented rapid fire spread in the narrow street. As firefighters moved in to fight the fire in Evans, the device at Graftons detonated less than 100 yards away.

On 7 July it was reported that residents of the high-rise flats in the New Lodge were pressing housing association officials to clamp down on fire raising, following a fire on the ninth floor of the Churchill House tower block, when vandals set fire to a number of tyres. Firefighters were in attendance within minutes and extinguished the fire but worried residents felt that only the quick action of the Brigade prevented what could have been a major tragedy. The fear was that fire could have blocked the escape route from the upper floors and lives could have been lost. At the time firefighters felt that this was one example of why the Brigade was valued in all of the areas in the city, despite the divisions brought about by the Troubles.

---

[233] Annual Report for the Fire Authority for Northern Ireland.

Early on 16 September one of the best-known bars in the city, the Old Vic in Little Donegall Street, was badly damaged by an accidental fire. Firefighters with five appliances used breathing apparatus to fight the fire from inside the premises, despite the thick smoke and intense heat. A second attack was made with a jet through the roof of the building and the fire was eventually extinguished. On 13 November the store holding the Christmas stock of toys for Dr Barnardos was badly damaged by fire. The fire was thought to be accidental but local people, helping themselves after firefighters had extinguished the fire and left the scene, looted the stock of toys. A second malicious fire broke out in the same premises later that day. Many people in the city donated to the charity following the fire and subsequent looting.

An Individual Rescue

Most of the time fire service activity relies heavily on teamwork. In fact it would be almost impossible to deal with most incidents the service attends without the mutual support and additional capability that teamwork brings. However, on rare occasions firefighters have no choice but to work alone, and it is at these times that their determination and courage come to the fore. One such example occurred on 29 December when my old friend, Station Officer Harry Moore, was called to an incident as a supervisory officer. Harry was at home when just before 5am he was turned out to a house alight in Breda Drive. As he lived close to the incident he arrived in his staff car before the fire appliances. When he arrived a neighbour had already broken down the door, the neighbour's daughter said:

> It was at this stage that a fireman arrived on his own. He shouted 'Where are they. Where are they?' and raced past me. He got the daughter out of the front room and was bringing her down when he was overcome by fumes. If it hadn't been for him, Mrs Keith's daughter would not have got out.[234]

Harry said:

> I just rushed upstairs to see what I could do. A neighbour told me some of the family were still in the house. I went to the front

---

[234] *Belfast Telegraph.*

room, which was on fire. I couldn't see Mrs Pollin because of the smoke, but I could hear her breathing across the room. I managed to get her out of the bedroom. She was in a semi-conscious state. I passed out myself for a few moments because of the fumes, but the police revived me.[235]

---

[235] *Belfast Telegraph.*

PART FOUR

1984 to 1989
The Long Road to Peace

1984

DURING THE EARLY years of the 1980s the pattern of the Troubles in Belfast began to change. There were fewer explosive devices being used, although there were still many incidents of this type, and more shooting incidents in the city, as the campaign of terror shifted somewhat to other parts of Northern Ireland and the mainland United Kingdom. In January 1984 it was agreed, reluctantly on the FBU's part, to end the practice of having a firefighter on watch room duty and Westland station was fully opened, with the introduction of a second appliance at the station. Operationally the Brigade had its lowest percentage of civil disturbance calls for several years.

Maysfield Leisure Centre

At about 1.30pm on Saturday 14 January a fire started in a room being used to store gymnastic and judo mats in the Maysfield Leisure Centre. Approximately 200 people were using the centre at the time. There was a judo competition and a man taking part opened the store door to bring out some of the mats that were there, he discovered the fire: 'I opened the store door and saw flames about five feet high,' he said.[236]

The fire developed rapidly and smoke began to emerge from the storeroom, spreading down the corridor, through openings and into adjoining rooms. The Brigade was called at 1.38pm, approximately three minutes after the fire had been discovered, and fire appliances from Central were in attendance within a minute of the call being received. As they arrived

---

[236] *Belfast Telegraph.*

Station Officer Maxi Carson saw smoke pouring out of the building, but he could see no flames.

In all, six people were killed in this incident. At the inquest into the deaths a police officer praised the Brigade saying, 'there was a calmness among the Fire Brigade as they went about their task. They were very brave men, I will never take them for granted again for their courage'.[237]

Gordon Latimer was on the first appliance that arrived at Maysfield. He remembers,

> One of my best friends, he found a mother of two children. She was caught up in five-a-side nets and as he went to move the nets the net caught his facemask on his breathing apparatus set and he took at least one good breath of thick, black, acrid smoke. I remember him telling me afterwards, it was like somebody punching you in the chest with a sledgehammer, the pain and the effect of it. He got his mask back on and they got out. It literally was so black that you could hold your hand up in front of your mask and not see it. I can still remember the bodies laid out in the foyer and firefighters just crying, you know, openly. Six people died if I remember rightly.[238]

Hundreds of people attended the funerals as investigations into the cause of the fire began. Assistant Chief Fire Officer Harry Boyce from the London Fire Brigade was appointed to look into the tragedy, with a reference to examine fire safety measures and to make recommendations to prevent a similar incident in the future. The initial forensic investigations into the cause of fire were inconclusive but it could be seen that it would be difficult to start a fire in the mats accidentally, as they had a tough outer covering. Eventually, the fire investigation indicated the most probable cause as direct ignition but there were no signs of an accelerant or any kind of incendiary device being used.

At that time fire safety legislation had not been updated in Northern Ireland in the same way as it had been in the rest of the United Kingdom. However on 28 March, in what appeared to be a reaction to the Maysfield fire, it was announced that fire prevention laws in Northern Ireland would

---

[237] *Belfast Telegraph*.
[238] Gordon Latimer.

be brought into line with those in the rest of the United Kingdom. This meant that the Fire Authority would have new powers to inspect many more public premises, comprising places of public entertainment, smaller hotels and guest houses, which could all now be issued with improvement notices and fire certificates.

On 19 January, in what seemed to be a copycat incident, a number of sports mats were set on fire at the College of Business Studies in Brunswick Street. Firefighters with four appliances quickly extinguished the fire and no one was injured.

Discrimination in the Brigade

On 2 July The Fair Employment Agency issued their report on the Brigade. They indicated that, even though the Fire Authority was not being accused of direct discrimination, the ratio of Protestants and Catholics in the Brigade was unbalanced, and the number of Catholics was 'substantially below the proportion to be expected'. The figures were reported as approximately 79 per cent Protestant and 18 per cent Catholic with the rest being others. The report also indicated that the Fire Authority had committed to change the proportions by introducing new employment procedures in selection. The new selection procedures meant that applicants would be given a number on application and this would be used to identify them during the written, manual dexterity and physical tests, with names only being disclosed at the interview stage. I had been appointed to the role of Station Officer on the flexible duty system (on call at evenings and weekends) in charge of the Training School the previous year, and was responsible, under the direction of my Divisional Officer Dessie Graham, for implementing the practical procedures for recruitment using the new policies. I was pleasantly surprised to find that if the numbering system was adhered to, the proportions of successful Catholic and Protestant applicants were the same as the proportions applying to join the Brigade. Over time this led to an improvement to the religious ratio in the Brigade.

Overnight on 8 August there were several serious street disturbances in the west of the city, to mark the 13th anniversary of internment. Bricks, petrol bombs and blast bombs were thrown and a bus was hijacked and set on fire. The security forces responded with rubber bullets. The disturbances lasted for eight hours and it was estimated that several hundred petrol bombs were thrown and 200 rubber bullets were fired. At about 3.22am on 9 August a number of petrol bombs were used to start a fire in the Whitehouse Dye Works. It took firefighters, with four appliances from Glengormley, Whitla and Westland using seven jets and a ground monitor, several hours to extinguish the fire. The warehouse and its contents were virtually destroyed.[239] Rioting broke out again on 12 August when police attempted to arrest a man at a rally in Andersonstown. The police used rubber bullets and a man was killed after being struck by one of them. About a dozen people were hurt. That night the rioting intensified and eight vehicles, including four buses, were hijacked and set on fire. The rioting, petrol bombing and burning continued over another ten days and nights in different parts of the city.

Just after 9.15pm on the night of 23 August, firefighters responded to a malicious fire in a furniture warehouse in an old church building in Nansen Street. Firefighters with eight appliances used six jets and two monitors from height appliances together with eight breathing apparatus sets to extinguish the fire. However the warehouse was badly damaged.[240] As the fierce fire engulfed the building a firefighter was injured following a small explosion, but he was released from hospital the following day. The bombing and burning continued.

Around that time the Brigade was having difficulties when attending incidents at Divis Flats. It wasn't unknown for children to throw things at the Brigade, but when they were being thrown from the roof of a block of flats things were getting pretty dangerous. At one incident someone actually threw a bathtub from the flats at a fire engine. The Station Commander based at Central at the time, Jackie Beattie, issued a personal plea through the press to the children and their parents to stop the practice. On the afternoon of 16 December, at about 2.50pm, there was a difficult accidental fire in the engine room of the English Star, a refrigeration vessel being built in the shipyard. The engine room was well alight when the Brigade arrived and 11 appliances attended, two fuel tanks exploded in the intense heat. Firefighters used four jets, one foam jet and a large number of breathing apparatus sets to extinguish the fire. Twenty-four firefighters involved in firefighting had precautionary medical check-ups at the Royal Victoria Hospital on their return from the incident.[241]

---

239 Annual Report for the Fire Authority for Northern Ireland.
240 Annual Report for the Fire Authority for Northern Ireland.
241 Annual Report for the Fire Authority for Northern Ireland

1985

THE MOST SIGNIFICANT political issue during 1985 was the Anglo-Irish Agreement, which gave the Republic a consultative role in the running of Northern Ireland. Unionists were totally opposed to this and they disrupted councils and other local government organisations, including the fire authority. The unionist opposition was so deep it lasted for years and resulted in political boycotts, mass rallies and an increase in loyalist violence. Chief Fire Officer Clive Halliday was appointed as Firemaster of Strathclyde Fire Brigade and he left to take up his post at the beginning of the year. Deputy Chief Officer Billy Beggs was appointed to take over.

The Troubles continued and at lunchtime on 14 June firefighters stood by as buildings in Gloucester Street were evacuated due to a warning of a van bomb. The device, in a white van, detonated as the ATO was arriving and a large part of the city centre was devastated, with buildings in Chichester Street suffering most damage. A number of people were injured by flying glass or suffered from shock. About 40 minutes later, firefighters, who were still working at the scene in Chichester Street, had to run from collapsing masonry, which fell from a building that was damaged in what may have been another blast. It was a lucky escape.

An Explosion Devastates Central

The fire station in Chichester Street was the headquarters of the Belfast Fire Brigade before amalgamation and it covered the city centre. Because of its central location the station's firefighters dealt with many of the bomb explosions during the Troubles. The location was a good one for covering the area but it did have one disadvantage in the Northern Ireland context, it was right next door to the Petty Sessions Court, near to the Recorder's Court and directly opposite the High Court. This meant that when the courts were attacked in terrorist operations, which were a frequent occurrence, the station often sustained collateral damage. On 29 July a van, which was packed with approximately 500 lbs of explosives, was left outside the Recorder's Court in Chichester Street. When the device detonated 30 minutes later the blast could be heard ten miles away. The Recorder's Court and the Petty Sessions Court were both badly damaged and buildings within a 500 yard radius were also damaged. Chichester Street Fire Station was badly affected, with ceilings being brought down and glass from windows, and the engine room doors, being blown into the street.

Louis Jones recalls:

> I became a station officer in 1985; it was on White Watch in Central. On the first night duty in charge as a Station Officer I remember hearing the screech of brakes and we looked out and this van had stopped outside. The next thing somebody says over the speakers, 'you have to evacuate the fire station, there's a bomb at the front of the station'. I thought, 'Right.' So we all sort off headed down. We took all four fire engines in a row; we drove them out of the fire station past the bomb. Whenever I look back on that I think it's ridiculous. But we drove out and drove round to the Laganbank Road and we parked there. Now it was probably the biggest one that hit the station because it completely destroyed the upper floors. It was decimated, the whole first and second floors were gone, the ceilings were down, everything was ruined.[242]

As does Chris Kerr:

> My first night duty was the immediate aftermath of the bomb, which had devastated Central Fire Station. So I came in to an operational fire station with ceiling tiles and dust on the floor. With the dormitories out of action, the galley was out of action, the emergency tender was running from Whitla Fire Station, the turntable ladder was running from Westland Fire Station and the two pumps were running from the yard in the old Central Fire Station. We lived and slept in what was the old training centre just across the way, it was undamaged and there were a few bunk beds and a little kitchen that was hastily adapted for our use.[243]

On the morning of 11 September an amusing (at least to firefighters) incident occurred – one, which could only happen in Northern Ireland – at the Forum Hotel in Great Victoria Street. One of the guests was eating his breakfast, having left his suitcase in the lobby of the hotel when an alarm clock went off in the case. The hotel was evacuated and a security guard carried the case into the street. The resulting bomb scare lasted for almost two hours and stopped rush hour traffic in the area. The incident was resolved by the ATO, using two controlled explosions to clear the suitcase, scattering the man's clothing over a wide area.

---

[242] Louis Jones.
[243] Chris Kerr.

On 6 November the Chief Fire Officer, Billy Beggs, approached Sinn Féin in an attempt to get their support to try to stop youths in the city attacking fire crews. Some young people in the Lower Falls area thought it was good fun to swing breezeblocks in plastic bags and throw them at fire engines as they were proceeding to fire calls in the area. Breezeblocks were also being dropped from Divis Flats onto the machines. That day, firefighters were raising funds for a Northern Ireland children's cancer charity by walking 20 miles in relay and wearing breathing apparatus. The project was initiated when it was discovered that one of the young relatives of a firefighter was dying from cancer. Station Officer Jimmy Hamilton said, 'We wanted to do something to help this needy cause. There just aren't the facilities for kids with cancer and this fund is trying to change that. We want to help it'.[244]

**

On 15 November the Prime Minister and the Taoiseach signed the Anglo-Irish Agreement. Unionist opposition to the agreement lasted for years and resulted in political boycotts, mass rallies and an increase in violence throughout the city. Once again there was a major increase in incidents for firefighters and the Brigade was particularly busy.

1986

BOTH UNIONISTS AND republicans were wrong-footed by the new phase of Anglo-Irish policy. Unionists varied between disorderly protests and boycott politics. But nationalist support for the agreement encouraged Sinn Féin to drop its traditional policy of not taking the seats it was elected to. In January the Anglo-Irish Agreement started to affect the fire authority when a member said that he would try to cancel a longstanding contractual arrangement to fight fires over the Irish border in Donegal. In addressing the issue the Chair, Dennis Connelly, criticised members who 'played politics with firemen's lives'. When the meeting took place the authority rejected any suggestion that it would not honour its agreement. The difficulties remained however, and in April a meeting of the Finance and Personnel Committee was brought to a halt by protesting unionist members, objecting to the Anglo-Irish Agreement. Operationally, after an unusually quiet January, opposition to

---

[244] *Belfast Telegraph.*

the agreement kept the Brigade busy as the bombing and burning intensified in February.

On 3 March, during a day of action over the Anglo-Irish Agreement, violence flared as a crowd rioted in Ballysillan, throwing stones and petrol bombs at the police. That night was one of the worst nights of violence seen in the city for some years. Barricades were raised in several areas and there was shooting, stoning and widespread petrol bombing as dozens of cars and shops were set on fire. The trouble continued through March and into April, with several police officers houses being petrol bombed as the bombing and shooting continued.

On 18 April, and into the next morning, the home of a retired police officer was petrol bombed in Sandy Row and cars and vans were hijacked and set on fire in the Shankill and Crumlin Road areas. The rioting, stoning and petrol bombing continued. On the morning of 20 April at about 7am, the Free Presbyterian Church in Clifton Park Avenue was damaged by fire. That night the home of a Catholic family was petrol bombed on the Stewartstown Road and the Windsor Superstore on the Donegall Road was badly damaged by fire. Firefighters arrived to deal with the Donegall Road incident as flames swept through the building and the roof started to collapse. However, a hostile crowd gathered and threw stones and petrol bombs at the firefighters, cutting their hoses as they attempted to extinguish the fire. This kind of event happened fairly regularly during the Troubles:

> I was driving at the time, I was operating the pump, so we had the hose in from the standpipe and this guy just came out and produced this big knife and he just stuck it in the hose. You just got loads of stuff like that, it was routine.[245]

> I can remember on the Castlereagh Road in my first couple of weeks operational, we were fighting a fire and I remember knives coming out and cutting the hose, and the crowd getting agitated.[246]

The Chief Officer attended the incident on the Donegall Road and condemned the attacks, but said firefighters had got used to such incidents and that 'the firemen just had to get on with the job'. [247]

---

[245] Nick Allaway.

[246] Ken Spence.

[247] *Belfast Telegraph*.

On the morning of 24 April, at about 4.30am, a fire broke out in a house in Millbank Park. The family of seven were in bed at the time and the mother managed to get out of the house. The father threw a baby from the upstairs front window to those standing below. He then threw another child out and jumped from the window himself, shouting to a neighbour that the other children were at the back of the house. The neighbour went round to the back. He said:

> Suddenly the place caught ablaze. I climbed onto a small roof ledge above the back door and shouted to the kids to come to the window. I could hear them crying. But the smoke was thick and I couldn't get in. I'm surprised that there was anyone alive at all.[248]

At that point the Brigade arrived and Firefighter Geoff Bannerman, wearing breathing apparatus, climbed into another bedroom that was engulfed in smoke. In the first bed he found a female child and handed her out to another firefighter. He then located two other children and brought them out as the fire was being extinguished. The three-month-old baby died from his injuries, his four-year-old sister, the first child rescued by firefighters, died from the effects of carbon monoxide poisoning and their father was seriously injured. Firefighters tried to resuscitate the children but to no avail. A woman at the scene said: 'They tried to resuscitate them with mouth-to-mouth, but they couldn't bring them round. Even one fireman was crying'.[249]

On 12 May four firefighters from the Blue Watch Springfield began a 123 mile walk around Lough Neagh to raise money for the Children's Cancer Unit of the Royal Victoria Hospital. The four, John Boyle, Nick Allaway, Robert Owen and Gerry Kelly raised £6,030 in all.

Martin Duffy: Death of a Firefighter III

Martin Duffy was a firefighter in Belfast, but he also had a part-time job as a taxi driver, not unknown for a member of the fire service, but somewhat more hazardous in Belfast. Just before 11.00pm on 19 July he went to pick up a fare at Chichester Park Central, just off the Antrim Road. Two girls gave him directions to the street and minutes later they heard five shots. They told the inquest that Martin had come stumbling past them, calling 'I'm shot, I'm

---

248 *Belfast Telegraph.*
249 *News Letter.*

shot, help me.'[250] He had been shot four times, in the right eye, chest, left arm and left elbow. He collapsed in the car park of the Chester Park Hotel and was treated by nurses who had been in the bar at the time. He died shortly after midnight. Martin was 28, came from Manor Street and had been in the Brigade for eight years. Firefighters from the city formed a guard of honour at his funeral and colleagues from Dublin, Cork, Limerick and Strathclyde attended. His brother, John, said:

> He didn't have a sectarian hair in his head. He would have gone anywhere. Only last week his station commander said he went beyond the call of duty to rescue a man from a house fire on the Shankill Road. He got the man out and saved him with mouth-to-mouth resuscitation. He was proud to be a fireman.[251]

The night of 8 August was one of street violence in the Falls Road area. There was automatic gunfire, and approximately 100 petrol bombs and several blast bombs were thrown at the security forces who responded with 107 rubber bullets. The trouble continued for another three nights. Chris Kerr remembers:

> One of the busiest night duties I had was on internment night in '86. We attended our first call just after 6.00pm in the evening and I think it was about six or half past six in the morning that we returned to station. There were over 30 calls that my own appliance attended ranging from property fire, shops on fire, portacabins, all by and large a direct result of arson or linked to the disturbances that were going on. And indeed firefighting amidst petrol bombs, blast bombs, gunfire directed at the security forces. We were firefighting just off the Falls Road in Clonnard; there was a row of houses under construction on fire. You could picture the scene, very little visibility with the smoke in the street, lights in the street had been rendered useless because of a desire to keep the area in darkness. I was on a jet with another, experienced firefighter, when he said, 'we have to go the situation has deteriorated', and I thought, 'that's interesting' because I presumed the people who were around me in the shadows were armed and that they were soldiers and it was only when we turned the water off and the

---

[250] McKittrick, Kelters, Feeney and Thornton (1999).
[251] *News Letter.*

light of the fire illuminated the fact that they weren't soldiers, there were a number of masked gunmen in the area.[252]

**

On 16 December a device of about 800 lbs of explosives, planted on a bus, detonated outside Lisburn Road Police Station. The station was reduced to a pile of burning rubble and almost 700 houses were damaged in the blast. Firefighters extinguished the fire despite the hazard of exploding bullets coming from the police station. A gas main, which was burning fiercely, was left to burn and Divisional Officer Jack Fell explained that it would take some time to extinguish the fire, which was still burning below the ground floor of the station. He said that the two gable walls of the police station could collapse at any time, and that it was best to allow the escaping gas to burn off, as any gas that hadn't burnt off, could lead to an explosion. This complicated the operation but they had 'floodlighting, ready to work through the night and well into tomorrow if necessary'.[253]

1987

THE UNIONIST PROTESTS against the Anglo-Irish Agreement affected the running of the fire authority in 1987 and in March four unionist councillors from Belfast resigned their positions as members. This followed similar action on the Education and Health Boards. The bombing, burning and street violence continued throughout the year.

Just after 10.00am on 13 March, two devices detonated in the new Smithfield Market within the space of ten minutes, and the premises were badly damaged. Two armed men planted the devices and a police officer was blown off his feet and slightly injured in one of the explosions. The new building had been due to open within weeks as a replacement for the old market, which had been destroyed by a fire caused by incendiary devices earlier in the Troubles. That evening two men were badly burned in a fire in a restaurant on the Lisburn Road. The restaurant was badly damaged in the incident. At about 4am on 18 March, the Poleglass Library and Youth Centre were badly damaged by fire, and late in the evening of 19 March, a blast bomb was thrown at Springfield Road Police Station causing minor damage.

---

[252] Chris Kerr.
[253] *Belfast Telegraph.*

On the evening of 1 April, about an hour before the Northern Ireland football team played England in the European Championships, a car bomb detonated 100 yards away from the rear entrance to Windsor Park Stadium. The blast started a fire in a nearby garage but due to a telephoned warning the area had been cleared and there were no injuries. The next morning a suspect device on the M1 Motorway caused traffic chaos during the rush hour. Overnight on 2 April, a fire was started by petrol bombs in a training centre in Beechmount Avenue. The centre was owned by the Catholic Church and was used to train both for the diocese and for SPRED (special religious training for the disabled). The ground floor was damaged in the attack. On 3 April, a prominent republican was shot dead by the UVF when he answered a knock at his front door. Three nights of rioting in the north and west of the city followed the killing. On 6 April, over 30 bomb scares, including suspected devices on 12 hijacked vehicles, brought the city to a standstill and caused the evacuation of the police station in Donegall Pass. The managing director of Citybus, Werner Heubeck, carried two devices off different buses. He boarded the buses, at Wellington Place and near the police station at New Barnsley, and removed a beer keg and a package before the security forces arrived. He had done a similar thing earlier in the Troubles, taking a suspect parcel off a bus. He said, 'I have done many things in my life and this was just one of them. It was not the first time'.[254]

At Paisley Park a device detonated on a bus, setting it on fire, and another device detonated at Kilwee Industrial Estate. A hijacked lorry was found outside Donegall Pass Police Station, but following two controlled explosions the ATO declared it a hoax. All of the bomb scares that day were found to be hoaxes apart from four, which involved actual devices. That night five security bases were attacked, three of them with blast bombs and two with mortars. No one was injured. Attempts were made over a number of days to hold the funeral of the republican who had been killed on the 3rd and, on the 7th and 8th of the month, those attempts brought more violence between the police and mourners. The 8 April saw a yet another night of violence in the city with stones and petrol bombs being thrown into the early hours of 9 April. Petrol bombs were thrown into Corry's timber yard on the Springfield Road but staff managed to extinguish the fire prior to the arrival of the Brigade. Thirty vehicles were hijacked and a bus was set on fire on the Oldpark Road. A fire appliance was stoned on route to

254 *Belfast Telegraph*.

an incident in Springfield Avenue and its windscreen was smashed. The rioting continued for another two nights with shooting, petrol bombing, blast bombing and stoning. On the following two nights, the 11th and 12th, shooting and rioting occurred in loyalist areas as protests over the Anglo-Irish Agreement continued. On the afternoon of 16 April, a suspected van bomb outside the Garrick Bar in Chichester Street was dealt with by the ATO using a controlled explosion.

Early on 6 May sporadic trouble broke out in the city prior to the funeral of an IRA member who had died when a blast bomb he was throwing at Springfield Road Police Station on the 2nd exploded prematurely. Petrol bombs were thrown and cars and buses were hijacked. The rioting continued after the funeral when crowds of people hijacked around 20 vehicles, including buses and lorries, and set them on fire.

Street Violence

The street disturbances were always dangerous, with the risk of injury or worse. However, by this stage of the Troubles, the violence on the streets seemed from a Brigade perspective to have become more vicious. Nonetheless firefighters always attended incidents when required, even in the most difficult of circumstances: 'You used to get stoned all the time'[255] with 'frequent bricks thrown through the windows'.[256] William McClay remembers:

> It was more a case of them fighting with each other whenever they got a chance, at the interfaces, so it was. But they turned on us very, very quickly, very easily and started stoning us. I was hit a whole lot of times with stones.[257]

At the beginning of the Troubles there had been a certain humorous or mischievous element to some of the street disturbances: 'It didn't seem to be so serious,' Walter Mason remembers,

> It was a 'chase me, chase me' and 'stone them, stone them' type of thing, but there was a bit of devilry to it. Most of the crowds... were good-natured towards us at that time; they didn't see us as a target.[258]

---

255 Nick Allaway.
256 Roger Dawson.
257 William McClay.
258 Walter Mason.

Often alcohol would be involved. Roger Dawson recalls, 'There were quite funny occasions when people threw full beer cans at us. It is incredible how the beer gets into the cab; it just vaporises and finds its way in'.[259]

One example of the devilry to be seen in a crowd occurred at the end of the '70s when I was mobilised with a two-appliance turnout from Ardoyne to a lorry on the Falls Road, which had been hijacked and set on fire. The cab of the lorry was burning when we arrived and there was a fairly large crowd around it, with some of them looting the refrigerated trailer and its contents of frozen food. As was the practice then we stopped the appliances a few hundred yards away, with an escape route should it prove necessary. I got out and, taking Marty Magill with me, walked up to see if we could convince the minders with the crowd to let us extinguish the fire, or if we should just let it burn itself out. As we walked down towards the lorry, and came into throwing range, some of the younger element in the crowd started to throw bags of frozen chips at us. Marty, who had a great sense of humour, shouted down at them; 'Hey, what's this, chips, fucking chips, we want scampi!'

The shout went back, 'Scampi, they want scampi!' So they rummaged about on the back of the lorry and the next thing they were throwing bags of scampi at us. Eventually we were allowed to put the fire out and when we got back to Ardoyne we had a very nice scampi and chip supper.

However, during the 1980s, particularly following the hunger strikes and loyalist opposition to the Anglo-Irish Agreement, things changed for the worse and it became 'a different story, an evil element came into it... you couldn't even send a man out to put a hydrant in by himself'.[260] Walter remembers that: 'Virtually everything that you went to, especially at certain times of the year, you knew that you had to be on your guard. It was just a matter then of driving as quickly as you could, sometimes you got away with it but the majority of times you didn't'.[261] William Hoey and Joe Sloan recall that 'nearly every time you went out you were caught up in it'[262] and that

---

259 Roger Dawson.

260 Walter Mason.

261 Murray Armstrong.

262 William Hoey.

'we got windows smashed loads of times. Four or five times in one night we lost a screen on the pump'.[263]

You could often tell when crowds of people were going to kick off and attack you, and experienced officers-in-charge would learn how to read the signs and withdraw before things got too difficult. But you could always be caught out.

**

On Friday 8 May eight IRA members were killed by the SAS as they were attacking a police station in Loughgall, County Armagh. That night street violence broke out in the north of Belfast with rioting and gunfire. Between 150 to 200 petrol bombs were thrown and the security forces responded with rubber bullets.

Gordon Latimer described the incident:

> Probably the worst pasting we got was the night of the shootings in Loughgall. I suppose the difference is that at bonfires and in the marching season you pretty much expect trouble, so you are ready for it. You approach incidents in a different way, to the point where people are carrying fire extinguishers and stuff in the cab because of petrol bombings. You are much more alert and aware of it and ready to take steps to move on if you need to. The difference with Loughgall was it was just an ordinary night with no reason to suspect anything, going about our business, getting a few calls, and unbeknown to us there had been a supposed attack on a police station in Loughgall. The security forces were lying in wait and shot a number of republican terrorists, but we didn't know any of that.
>
> The first we knew about it was when we got a call to Dunville Street off the Falls Road. We drove up the Grosvenor Road and turned right onto the Falls Road and a huge crowd of people, which obviously wasn't letting us through, confronted us. We hadn't expected this. As with most of these things there was a spokesman at the front telling us to back off. The number

---

[263] Joe Sloan.

two appliance was right at the back of us, so we couldn't reverse. The driver of the second appliance was wondering why we were not going anywhere and obviously there was a heightened state of tension. The guy who was the figurehead of the rioting who was telling us to move back, and, almost like a gangster film, reached inside and pulled a revolver out and pointed it at the officer-in-charge, Geordy Goodman, who in fairness to him, in all calmness, made it clear to the guy through signals he wasn't threatening in any manner. He got out of the appliance and explained to this guy what the problem was and said, 'that's fine we are going but calm down.' The second appliance reversed back and started down the Grosvenor Road. We reversed back and followed them down. At that point we were thinking, 'god that was a close shave,' when a brick then came through the front windscreen from somebody who had stepped out of the entrance to the Royal Victoria Hospital. We went back to the station, got a spare appliance, changed over and got ourselves back on the run as quickly as possible. And away we went again.[264]

Lucky Escapes

On 27 October a number of masked and armed men planted four devices, each comprising a five-gallon can of petrol with explosives attached, at Mackie's Factory on the Springfield Road. Fifteen minutes later one of the devices detonated starting a fire. The other devices were defused by the ATO using controlled explosions. Four appliances stood by to deal with the incident. Later that day a booby-trap device detonated nearby injuring a soldier and a 66-year-old woman. Neither of them was seriously injured. It was claimed that this device had been intended for security forces that attended the earlier incident in Mackie's. A number of firefighters had a lucky escape:

we were standing by nearby. I sat on a low wall on the other side of the road... we must have sat there for the best part of an hour, not able to do anything while we waited on the building being cleared. Eventually we were called to another situation, but whilst we were away a soldier in a passing foot patrol at the site

---

[264] Gordon Latimer.

at Mackie's was injured when a grenade exploded behind the wall that we had been sitting on. So we were pretty sure that the whole time that we were sitting on that wall there was a grenade sitting right behind it ready to be detonated at any time.[265]

This was by no means the only time firefighters were lucky to escape injury or worse, and we would often say we were lucky or 'someone up there was looking after us'. Joe Sloan remembers an incident in Duncairn Gardens where one of the team kicked an object [as they arrived on the scene] and then 'found it was a live mortar that hadn't gone off. And then it was sort of get out of here!'[266]

Gordon McKee, William McClay and Kenneth McLaughlin remember particularly lucky escapes:

In Great Victoria Street, a bomb had gone off on the buildings to the left hand side and it was blazing. We went down and set in ground monitors and when we looked round, on a building opposite, another big bomb was hanging on a grille. So there was Joe Love, myself, I think Harry Welsh was at that, quite a few of us ran up the street again and just as we got up to the top of the street the bomb went off. Geordie Roundtree was the LF, he got blew across the top of the TL, Joe Love got blew underneath the TL and I managed to get round the front of it. So if we hadn't have moved just when we did, when somebody noticed that the bomb was there and we ran like hell, we would have been well mangled.[267]

One of the boys walked over to a lamppost and called the boss over and said; 'there is something tied to that lamp' and we looked over and there was something taped on it. It was a trap, a booby-trap for the army or police.[268]

There was a 500 lb bomb in a van that had crashed on the Antrim Road. The main detonator hadn't gone off and the army were happy enough that we washed out all this fertiliser. So Dorky and me took shelter behind this wall and we had the jet and we were

---

265 Brian McClintock.

266 Joe Sloan.

267 Gordon McKee.

268 William McClay.

hosing out the thing. We said to the bomb disposal guy, 'everything OK?' 'Oh sound as a pound, it's only powder.' The next thing clink, clink, four mortars, we had washed them onto the road.[269]

The Troubles continued to the end of the year. On the night of 3 November a fire badly damaged a first floor storage area in an Argos shop in Royal Avenue. Early on 10 November, Derriaghy Cricket Club was badly damaged when someone climbed a security fence, poured petrol on the building and set it on fire. In the late afternoon and well into the night of 25 November, a series of bombs and suspect devices caused havoc in the city. Hijacked vehicles were left outside ten security bases and police patrols came under gun, petrol bomb and stone attacks, in different parts of the west of the city. Three incendiary devices detonated in cars outside police stations in Donegall Pass, North Howard Street and Lanark Way. The same day a number of men smashed the window of the guards van of a train in Central Railway Station and threw in two incendiaries. They also left a third device on the platform as they made their getaway. The devices in the train detonated and badly damaged the carriage. The third device was defused by the ATO. In addition, an incendiary device badly damaged a bus, there was rioting and shooting and the security forces responded with rubber bullets.

1988

TALKS BETWEEN THE SDLP leader, John Hume, and Sinn Féin leader, Gerry Adams, began in January. They were highly controversial at the time but they would eventually lead to the peace process of the 1990s. However, the year was marked by a series of savage killings and the rioting, bombing and burning continued throughout the year. The Chief Fire Officer, Billy Beggs, retired at the end of the year and Ken McNeil, who had been in the service since 1959, was appointed as the new Chief Fire Officer for Northern Ireland.

Early in the evening of 9 January a 500 lb car bomb was left outside the magistrates' court in Chichester Street. A warning was received and firefighters in Central Fire Station, next door to the courts, drove their appliances out of the station and past the car to safety. The device detonated at approximately 7.00pm, badly damaging buildings in the street, including Central. A spokesperson for the Fire Authority said that although there was 'blast damage throughout the building' all the equipment had been saved

---

[269] Kenneth McLaughlin.

and 'all the services were back to normal immediately after the explosion'.[270] On 24 February a group of soldiers arrived to close and lock the security gates in the city centre. As they did so a 200 lb booby-trap bomb hidden behind wooden hoarding in Castle Court was detonated, it killed two soldiers and caused a massive amount of damage to buildings in the area. A second device, designed to catch anyone involved in a follow up operation, failed to explode and was defused by the ATO; the Brigade attended and assisted with body recovery as well as making the affected buildings safe.[271]

A Deadly Sequence of Events

On 16 March, three men were killed in an unprecedented attack at Milltown Cemetery. A gunman opened fire on thousands of mourners attending the funeral of three IRA members who had been killed in Gibraltar earlier in the month. All three men were thought to have died while pursuing the gunman from the scene of the attack. The crowd eventually caught the gunman and beat him unconscious, before the police arrived. Serious street violence erupted in several parts of the city following the Milltown cemetery killings, hundreds of petrol bombs, stones and at least one acid bomb were thrown, many vehicles were hijacked and burnt and several houses were attacked and petrol bombed. Shooting could be heard in many parts of the city and the security forces responded with rubber bullets. The Brigade was kept very busy. On 19 March, two soldiers in plain-clothes and in an unmarked car drove into the cortege of the funeral of one of the people killed at Milltown Cemetery. The two corporals, said by the army to be technicians engaged in routine communications and radio work, but thought by local people to have been involved in some sort of undercover operation, (neither version really explains how they got caught up in the funeral) were pulled from the car and punched and kicked to the ground. They were driven 200 yards to Penny Lane where they were shot several times. Most of this savagery was caught on the surveillance camera of a military helicopter and shown on television around the world.[272] The fire service was called to deal with the aftermath:

> The two corporals that got killed in west Belfast, a truly horrible incident, when they were basically set upon by this mob. One of

---

[270] Annual Report for the Fire Authority for Northern Ireland.
[271] Annual Report for the Fire Authority for Northern Ireland.
[272] McKittrick, Kelters, Feeney and Thornton (1999).

the guys pulled out a revolver put a few shots into the air but eventually was overwhelmed. Somehow these guys had managed to get themselves into the middle of this funeral. There was a lot of paranoia about and these guys were spotted almost instantly. The upshot of it was, the car that they were in was taken away to Andersonstown, and at that point we got called out because they burnt it. You have to be careful with cars, especially if they have been burning for a long time and there's a picture of me somewhere getting a bit too close to this car and putting it out and the whole thing went up like a massive firework. And I tell you what, I went back very, very quickly.[273]

On 21 April a workman was shot on the roof of the law courts in Chichester Street by a sniper using an automatic weapon. Despite the obvious danger from further shooting, firefighters lowered the wounded man, who they tied into a stretcher, to street level using a turntable ladder. On 7 July a booby-trap bomb outside the Falls Road Swimming Pool detonated, killing a 24-year-old man and a 60-year-old woman. The woman was still conscious when she was put into an ambulance but she died later in hospital. A four-year-old child and her grandmother were injured in the explosion. The grandmother said:

> There was a big bang. It seemed as if the wall came out and my granddaughter just started to scream. Ciara's cardigan was just full of blood and it was terrible.[274]

A hostile crowd prevented army medical teams from reaching the injured civilians and an ATO was killed when he detonated a booby-trap bomb, by stepping on a pressure plate during the follow up operation.

A wedding prank very nearly went badly wrong on 12 August as the bride was tied to railings at the law courts in Chichester Street. A van bomb was left across the road as she was being tied up and the driver of the van shouted a warning to passers-by. In the rush to evacuate nearby premises the struggling bride-to-be was overlooked and left helpless. However, firefighter Eddie Campbell came from the fire station across the road and quickly untied her, just minutes before the device detonated. Divisional Officer Mick Malone said: 'He took a chance but what else could you do.'[275]

---

[273] Nick Allaway.

[274] McKittrick, Kelters, Feeney and Thornton (1999).

[275] *Belfast Telegraph*.

The violence continued through August and intensified towards the end of the month. By 1 September 106 vehicles had been hijacked and set on fire. Long queues of commuter traffic snaked through massive roadblocks comprising the smoking black shells of commercial vehicles and private cars, and there were four major burning barriers on the Falls Road alone.

1989

IN NOVEMBER THE new Secretary of State, Peter Brooke, suggested that the government would be 'flexible and imaginative' if the IRA renounced violence. However, despite this intervention the Troubles continued throughout the year.

On 2 January New Year bargain hunters were caught up in a fire at the Co-op. As shoppers arrived at the front door they were showered with soot from a cloud of black smoke, belching from the roof of the Orpheus building next door, the plume of smoke could be seen all over the city. The fire had started in the boiler house of the former Orpheus Ballroom, which was now the New University of Ulster's art college annex. Four fire appliances were quickly at the scene and a Brigade spokesperson said there was no immediate danger to customers or staff at the Co-op.[276] No one was injured and firefighters, unable to access the college through the former Co-op food hall, deployed a turntable ladder to reach the roof and tackle the fire.

Held at Gunpoint

On the evening of 1 February firefighters in Central Fire Station in Chichester Street were held at gunpoint, while an explosive device was thrown over the back wall of the fire station towards Townhall Street Police Station. As a follow up operation was being carried out a second device, consisting of a kilo of Semtex attached to a lamppost, detonated in nearby Oxford Street, but no one was injured.

While men with guns prevented firefighters from attending or dealing with incidents fairly regularly during the Troubles, this was an infrequent occurrence when firefighters were not at an incident. One example occurred earlier in the Troubles when a fire prevention officer was examining a defective chimney flue in Servia Street on behalf of the council:

---

[276] *Belfast Telegraph.*

There was nobody about at all to go so I took Slim with me, the waterman, and up to Servia Street. So we are sitting anyway and Slim's out in the car and I just got the wee dye pellet into the fire to see where the green smoke was going. The next thing this boy comes in with a gun! 'Come with me, come with me'. I said, 'I'm not going with you! I'm here to find out if this woman has a faulty chimney flue or not'. 'Come with me now!' The men threw bags over us and then took us round to a house; they were going to shoot us. It was in the Lower Falls and they interrogated us, fired shots beside our ear. They said we were Special Branch and they were going to shoot us.

My father-in-law worked in Ross's Mill up in Odessa Street, the linen mill, so I said to them, 'Go up there and he'll tell you'. The only thing that saved me was that they asked Alex 'Does he speak funny?' 'Well to you and me', Alex replied, 'He speaks funny for he never lost his Derry accent you see'. And that was it and he says, 'You just saved a man's life'. They were letting me go but they weren't letting Slim go, and I says, 'Well look boys I'll tell you something, its up to yourselves what you want to do'. I was shiteing myself. I said, 'You keep him and harm him and there will never be another fire engine in the Lower Falls and I mean that for I'm a union official'. So they let Slim go too, into the car and we went back to the station.[277]

Booby-Trap Car Bombs

As the Troubles continued through the 1980s the tactic of attaching a booby-trap device underneath the car of a member of the security forces, detonated by a mercury tilt switch, became a favoured weapon of the bombers. While the Brigade was not always called to this type of incident, we were required if an injured person or a body had to be cut from the wreckage, or if the car had caught fire. On 27 February, just before 2.00pm, a booby-trap device detonated under a car at the junction of North Road and the Upper Newtownards Road:

We had just finished lunch; we heard a very loud explosion in very close proximity to the fire station. Our Station Officer opened

---

[277] Jimmy Armstrong.

the front door of the fire station, could see some smoke, heard some screaming and we immediately had a running call for the pump to what was a car on fire. The head cloth was on fire and I always remember a school-crossing patrolman shouting at me 'He's dead, he's dead!' A number of vehicles had been damaged by the blast.

We extinguished a very small fire in the vehicle and realised that there had been a casualty, the blast had blown up from the underside of the vehicle, under the driver's seat. I managed to remove the casualty from the car; he was half blown out of the car onto the ground. Because the vehicle had blown onto the footpath and into the hedge, I checked below it in case there was a child pedestrian and was confronted by the remains of what I thought was part of the device. Very quickly we were joined by our second crew and a colleague came down to assist me and a plain-clothes police officer came down. The person was still living so we removed him. I still remember having to have a separate blanket to put intestines and things in, even though the person was still living, onto the stretcher and we carried the stretcher down to the ambulance and left the scene. I suppose the other abiding memory of that was going back to the station and washing body fluids off the old yellow leggings etc. But we went back on operational duty; the appliance was available. I suppose the normality in abnormality. It transpired it was a retired police officer and he died on arrival at hospital.[278]

A Fight in Springfield

On the evening of 3 June a fire officer and another person were taken to hospital following a fight at Springfield Road Fire Station. The row blew up after the Brigade had attended a fire in a derelict house in Springhill Crescent and afterwards a number of residents went to the fire station. The two people that were injured were taken to the Royal Victoria Hospital but they were not seriously hurt and were quickly released. A Brigade spokesperson said that there had been a slight misunderstanding about a dog being knocked down and the police were investigating the incident.

---

[278] Chris Kerr.

The pump got a call and unfortunately they hit a dog. They proceeded on, I remember the driver telling me that as soon as they got back they reported it to the police. They were on the way to a fire call and they couldn't stop. There were no markings or anything on the dog. A couple of hours later the owners turned up and unfortunately, rather than being ushered into a room and spoken to they got up the stairs, and said, 'Who was driving the machine'? The driver said, 'Well I was driving' and it started from there. There were fists, it started and it just progressed down into the engine room and out onto the street. Then we had the factions taking place and away we go again, all over the incident of a dog. It got out of hand, that's all, it got out of hand.[279]

On 9 June, in an anonymous letter to the *Belfast Telegraph*, a firefighter said:

What occurred that night was nothing short of a premeditated vicious assault on firemen by a gang of local thugs. The firemen to their credit chose not to raise one finger in their own defence, despite the fact that one of their comrades was hospitalised. Indeed, when a rival gang emerged on the scene and a minor riot ensued, one of those who had come to abuse the firemen was left at the roadside unconscious and injured while his associates ran off. It was left for the same firemen who had been threatened and beaten to take him in, care for him and arrange for his removal to hospital. Ironic? The incident at Springfield was not isolated. It was not a one off. It's just that rarely are these events brought to our attention. Go into any station, you can see for yourself the dents and holes left by bricks and rocks where fire engines have been used for target practice.

This wasn't the only time that a street fight affected Springfield station:

On one particular occasion three chaps came down drunk, said a few things, the opposing side saw them and attacked them out on the street, and then it slowly massed into a great group of people slowly working their way in towards the station. We were just getting a station refurbishment and there were all these fluorescent tubes ready for disposal, these were going all over the place, being used as javelins and weapons. The outer doors were

---

[279] Gordon Galbraith.

never locked as such, it was the inner portion that was locked and they managed to get in, get right in.[280]

Bonfire Night

The tradition in loyalist areas of the city (and the rest of Northern Ireland) is to light a bonfire on the 'Eleventh Night', the eve of 12 July, the biggest day in the Orange marching season. This was always one of the busiest nights for the Brigade and we were often involved at incidents over the period:

> You always had trouble just because of the bonfires. Most of it was good natured, I mean they used to put the bonfires near to people's houses and people's gutters just used to melt and they would call out the Fire Brigade. We used to have to try and make it clear to them, we're not putting out your bonfire, what we are going to do here is we are going to spray this woman's wall and we are going to spray her guttering, because at the end of the day it's damaging her property and we have to do something about it.[281]

> I got a punch in the mouth in Sandy Row that night. We turned out, I jumped out of the fire appliance and was immediately confronted by six drunk bonfire supporters who thought we were there to put their fire out, and he punched me in the mouth. We then explained to them that we weren't here to put their bonfire out, we wanted to put some cooling water on that house that was smoking quite rapidly next door, and they calmed down and we calmed down, and we put some water on the house and then we disappeared again.[282]

With competition between different areas it was common for bonfire sites to be guarded against rivals plundering bonfire material. On the night of 7 July, tensions were high in the New Lodge and Tigers Bay areas after a man, who was guarding a bonfire in the Tigers Bay area, was killed by four Catholic youths, one of them carrying a hatchet, who were apparently trying to seize loyalist flags as tribal trophies. That night rival gangs from the New Lodge

---

[280] Gordon Galbraith.

[281] Nick Allaway.

[282] Colin Lammy.

and Tigers Bay fought each other in street rioting for over three hours. On 11 July night itself there was sporadic rioting in different parts of the city. There were petrol bomb and gun attacks and once again the Brigade was very busy.

On the morning of 31 July, at about 9.50am, two young men were forced at gunpoint to drive a laundry van with a 1,000 lb. bomb on board to the High Court in Chichester Street. They shouted a warning and the area was evacuated before the device detonated at the court's main entrance, 45 minutes later. The explosion caused considerable damage over a wide area, including Ann Street, Victoria Street, Victoria Square and throughout the Cornmarket area. The city centre shook and the fire station took the brunt of the explosion, but remarkably no one was injured and only three people were suffering from shock. Debris from the van was blown over a wide area and a piece of it went through the back window of a car in Ann Street, more than a quarter of a mile away. A fire started in a car in the car park of the court and more than two-dozen cars were destroyed.

> I was standing in the engine room in Central and the doorbell rang and I was the only one about the place and I walked over to the door, put up the door and there was a guy, a security fella, standing across from me. He said, 'They have just put a big bomb in the courthouse, it's in that laundry van.' And when I looked across there was a laundry van sitting there. So, I ran into the station, I told them there is a bomb in the courthouse across the road, the four appliances were sitting in the engine room. We had to go out the back gate and we ended up walking right round to Custom House Square. And the bomb went up and there were Initial towels landing at our feet, in Custom House Square.[283]

The sad thing was that this had become almost routine for the firefighters of the city. Little did we know that even though attempts were being made to bring about some sort of peace process, things would get worse before they would get any better.

---

[283] Gordon McKee.

PART FIVE

1990 to 1994
Things Get Worse Before
They Get Better

1990

RELATIONSHIPS BETWEEN THE government and unionists gradually emerged from the post Anglo-Irish Agreement difficulties with talks about talks taking place, and discussions continued between the minister and Sinn Féin regarding conditions attached to peace talks.

For the Brigade, work to replace Central Fire Station was ongoing, however firefighters were becoming even more concerned about their safety in the station, and on 10 September the following letter from an anonymous firefighter was published in the *Belfast Telegraph*:

> Central Fire Station is the most bombed fire station in the world. Since July 1985, six bomb attacks on the adjacent courts and police station have also seriously damaged the fire station. There have been scores of hoaxes and false alarms. Despite this the station has never closed. The situation has become increasingly intolerable as Lower Chichester Street has taken on all the trappings of a high security complex. Recent gun attacks on police officers manning the security gates have sent firemen diving for cover. Invariably such incidents occur late at night and only firemen and security forces are in the area. Unlike the police we have no bulletproof jackets. Nor would we want them. Unlike the courts, the fire station is not fortified. Nor would we want it to be. The courts complex has become a battle of wills between the security forces and the IRA. It takes little imagination to realise that attacks on Lower Chichester Street will continue. Sooner or later a fireman could become another 'statistic'. The only solution

is to move to the proposed new Bankmore Street station as soon as possible. As the completion date is two years distant, a greatly speeded up construction programme is essential. To effect this it is likely that funding over and above the Fire Authority budget would be required. Senior fire service officers and indeed the Fire Authority are sympathetic to the plight of firemen serving at Central. But there are limits to what they can do at present, given financial constraints. Ultimately the only source of extra funding is government, in this case the NIO (Northern Ireland Office) I don't think firemen at Central are being unreasonable in asking for additional monies to be made available. I hope the Minister will agree.

Late on the night of 11 May a man who was visiting a woman in Daphne Street went outside into the street, poured a gallon of petrol over himself and set it on fire. The Brigade was in attendance in minutes but it was too late to save the man and he died of his injuries. On 28 May, a 33 tonne crane fell onto an estate agent's office in Arthur Lane, bringing down large slabs of concrete and masonry. Two people sitting at their desks were injured and initially it wasn't known if other people were trapped in the building. The premises were badly damaged and it took firefighters some time to search what remained of the first floor of the premises, making sure that everyone was accounted for.

On 18 August, as the bombing and burning continued, a number of people broke into an engineering company off the Newtownards Road and started a fire. The premises were badly damaged before the Brigade could extinguish it. At that time, it had to be established that three or more people had taken part in a Troubles-related incident, before compensation could be claimed from the Northern Ireland Office. This incident occurred at roughly 2.00am and was called in by a passer-by, so it wasn't clear how many people were involved. Therefore, the owner offered a £500 reward to anyone who could say how many people were involved in setting the fire. I was involved in a similar legal debate when I was a Station Officer in Ardoyne, it could be established that a group of people were outside a shop some hours before it was set on fire. I estimated that the fire had been burning for less time than that before we were called, and therefore the legal argument was about whether or not the fire could have been burning for several hours before being reported, and consequently could have been started by the group of people who were there earlier. I spent an interesting couple of hours being cross-examined on the witness stand, before the judge accepted my version of events.

Joy Riding

During the Troubles a craze for stealing cars to speed around the city grew amongst groups of young men. These so-called joy rides often ended badly as the cars crashed or were challenged by the security forces:

> At weekends you went to loads of cars on fire because there was a big problem at the time with so-called joy riders. What they would do is, for two or three hours just spin around in these cars and then just burn them in the west, at which point the police would call us out or somebody would call us out and we would put them out.[284]

On 30 September two joy riders were shot dead by the army, and a soldier was imprisoned for life for the murder of one of them. He was released from prison five years later and this decision led to serious rioting in nationalist areas of the city with 200 vehicles being set on fire.[285]

> There was rioting all over the place. We were coming on the night shift, it looked like we wouldn't be able to get into the station but we improvised and eventually the two watches changed over around 6.00pm as usual. Just above Springfield fire station at the junction with the West Circular Road, it's now a roundabout, but at that time it wasn't it was a T-junction. An old Belfast Gas transit van had been abandoned with a beer keg in the front seat. So this was literally 50 yards from the station gates. Sometime later we were standing up in the recreation area upstairs in Springfield looking out the front window to this scene of activity. Now why we were doing that with this potential device only 50 yards away was another question to be asked, but there we were. The army bomb disposal guys appeared in due course. And I remember this guy walking up in all his gear from a position below the station, walking up the white line in the middle of the road towards this van with the beer keg wired up sitting in the front seat; the

---

[284] Nick Allaway.

[285] McKittrick, Kelters, Feeney and Thornton (1999).

bomb disposal guy. And as he passed the station he looked up and looked directly at me and winked as he went past! Those are the really brave guys. So he went up and set his controlled device after inspecting the van. At that point on the Springfield Road there is a slight slope and a bit of a camber towards the station where the station gates were. Anyway, the controlled charge blew the hand brake off the van, we had been evacuated to the rear of the station prior to the controlled explosion but as soon as we heard the bang everybody of course rushed to the front of the station again to see what was happening. And just as we reached the front of the station the van started to roll down the hill, beer keg still on the front seat, and follow the camber into the station. So we beat a hasty retreat to the back again. Luckily it turned out to be an elaborate hoax.[286]

1991

POLITICAL TALKS, AIMED at reaching a peace deal, were continued in 1991. However, during the year and the following two, despite, or maybe because of the talk of peace, the violence increased. In better news, and in a long overdue event, Northern Ireland's first female firefighter, Heather Lysk made history by completing her training at Westland on 24 May.

A Major Restructuring: Belfast Becomes a Division

Work on a wide-ranging review of fire cover had been continuing since Chief Officer Halliday announced it in 1982 and, working to the Deputy Chief, Stephen Walker, John McClellan had finalised the planning for a major re-structuring of the Brigade. On John's promotion, to head up the training department, I was appointed as Brigade Projects Officer, with the responsibility of coordinating the implementation of the plans, together with three very competent people, Hazel Kelly the Project Officer, Station Officer Wesley Currie and Project Assistant Joy Chambers. The intention was to increase the number of divisions from five to six, re-draw the divisional boundaries and move some of the divisional headquarters. This would require a substantial change to the Brigade mobilising system to accommodate all of the changes in structure, including a revision of the call

---

[286] Brian McClintock.

sign allocation for fire appliances and officers. As part of the overall change Belfast would become a single Greater Belfast Division, which would include Glengormley station (A Division), and a new Belfast HQ was being built in Bankmore Street to replace the Chichester Street site. Over the previous two years between £4m and £5m had been spent on constructing new offices and stations as part of the programme. These changes went live at 9am on 3 June 1991 and this major change was effected 'without any implementation difficulties'.[287]

As a result, Belfast, which had been split between two divisions as a consequence of the amalgamation of the two Brigades in 1973, was recombined into a new city division with seven stations, as it now included the station at Glengormley. Over the reporting year the new A Division carried out 31 per cent of the operational workload of the Brigade, including 2,070 major fires, 3,388 small fires and 570 chimney fires, with 223 special services. All together the division attended 9,386 incidents.[288]

Even though this structural change took place with no difficulties, the industrial relations landscape continued to cause problems. Throughout the Troubles staffing levels of appliances in Northern Ireland exceeded those of most Brigades in the United Kingdom. The recommended staffing levels under the National Joint Council were five on the first pumping appliance and four on the second, on 75 per cent of occasions, with a crew of two on height appliances such as turntable ladders or hydraulic platforms. Northern Ireland usually rode five, five and three, in recognition of the additional difficulties of maintaining a service during the Troubles. As a result of the reorganisation it had been suggested by Her Majesty's Inspector of Fire Brigades that this situation should be reconsidered when the plans were being implemented. This soon became an assumption by firefighters, not unnaturally, that there was a desire to save money by reducing the crewing levels of the Brigade. On 4 June firefighters lobbied the Fire Authority and handed in a 1,000 signature petition, protesting against discussions on reducing the crewing levels of appliances. Ken Cameron, the General Secretary of the Fire Brigades Union, argued that the suggested reduction would affect the health and safety of both firefighters and the public. The meeting was stormy and three members of the Fire Authority

---

[287] Annual Report for the Fire Authority for Northern Ireland.
[288] Annual Report for the Fire Authority for Northern Ireland.

walked out in protest over the plans, however, nothing was decided and in October the Fire Authority indicated that they were not planning cuts to firefighter jobs.

On 5 January, as the Troubles continued, there was a concerted series of incendiary device attacks on shops in Northern Ireland, causing tens of millions of pounds worth of damage. The Sprucefield complex, the MFI store in Newtownabbey and the Castle Court centre in the city were among the main targets. Two hundred firefighters, many of them from the city, dealt with the various incidents, which were all made worse as high winds fanned the flames. The Sprucefield complex was badly damaged, as were the Locksley furniture shop in Dunmurry and the Lee Jeans factory in Newtownards:

> When I came down to Newtownards as a District Officer I got the Lee Apparel factory shop there with Newtownards station. And making pumps six and Controls coming on saying, 'sorry we can't send you any more pumps, they are all at Sprucefield' and Jack Fell and the rest of the Division were at Marks and Spencer's. Sprucefield was gutted the same night.[289]

Shops in Lisburn and Glengormley were badly damaged and a number of shops in the Castle Court complex were also damaged by fire. Things could have been worse but firefighters worked extremely hard at all of the incidents, over nine hours in all, to extinguish the fires with jets of water and prevent fire spread to other premises.

The shooting continued and several gun attacks took place in Chichester Street. On 24 January, it was reported that these attacks were taking their toll on firefighters based in Central Fire Station. An unnamed firefighter, quoted in the *Belfast Telegraph*, stated that he and his colleagues regularly had to dive for cover during gun and bomb attacks on the High Court building in Chichester Street. He produced a record indicating that there had been ten gun and bomb attacks in the Chichester Street area in the previous six months and claimed that firefighters in the station were now suffering from stress. It was argued that the long term solution was to expedite the move to the new station, which was being built in Bankmore Street at the time, but in the meantime the provision of bullet proof glass in the existing station

---

[289] Colin Lammy.

should be a priority for the Fire Authority. Shots had been fired in Chichester Street the previous night and the crew returning from a call in the emergency vehicle had to 'dive for cover during the attack'. Fortunately the driver saw 'flashes of gunfire in an entry and slammed on the brakes, but the appliance narrowly missed a lamppost. Inside the station 40 people who were attending a union meeting had to dive for cover when the gunfire started'.

The station was out of action for about an hour while the security forces cleared the area. Senior Divisional Officer Jack Fell, quoted in the *Belfast Telegraph*, stressed that the Fire Authority were very concerned about the situation:

> The safety of all personnel is our main priority and to that end there are on-going negotiations between principal officers in the Brigade and representatives of the Fire Brigades Union to see how the situation can be improved.

He said that principal officers of the Brigade were consulting with the contractors of the new station to see if construction could be brought forward. He was also conscious of the fact that the station could not respond to calls when a terrorist attack was taking place. 'This can lead to a delay of several minutes if another station is covering, depending on where the fire is. Every second counts in the case of a fire at a residential address where people may be trapped'.

The bombing and burning continued and over the night of 14 March a number of incendiary devices detonated in different premises in the city. In the worst incident a fierce fire took hold of Robinson's Bar in Great Victoria Street, with thick, black smoke rising into the air. Firefighters from four stations fought the fire using jets of water and height appliances. The first and second floors of the three-storey, traditionally built building were virtually destroyed, and it was thought that the cost of the damage would be in the region of a million pounds. When firefighters arrived at about 4.00am on 15 March, the fire had a good hold and flames were coming out of the front windows of the bar. Assistant Divisional Officer Bobby Pollock said that the crews had coped well in dangerous conditions to prevent the fire from spreading to adjoining properties:

> It was a good stop by the Brigade. There was a lot of heat and smoke and conditions were very trying. I am very proud of the men as the heat was phenomenal. We don't know at this stage

how many devices were planted, but the first and second floors are totally destroyed. The ground floor has managed to escape with water damage.[290]

Robinson's was on fire, they had fire bombed them as well, so we got thrown into that, and my job was to get up the stairs to fight the fire. We went in and it wasn't warm at all, breathing apparatus on, and you go up the stairs and it just gets warmer and warmer and in the end there is the old thing about your ears. If your ears start to burn it's time to go! And my ears were starting to crisp, you know, and I was starting to think this is very hot in here. Just at that point, they had put either a hydraulic platform or a turntable ladder up to the top and they had hacked a big hole into the roof and they decided to stick a hell of a lot of water through there, so that forced all the heat and the cinders and all the crap down our way. We just said it's time to go and we went down the stairs. We were out all night at that one. It was kind of nice getting back to the station and getting a shower after that; we were soaked through and absolutely exhausted.[291]

Six buildings were badly damaged by fire that night. On the afternoon of 27 February three young 'joy riders' died when the stolen car they were travelling in collided with a parked car and crashed into a lorry on the Upper Springfield Road. The driver of the car was 15, another 15-year-old died in the front seat and a 16-year-old in the back seat. It took firefighters over an hour to cut the bodies from the wreckage of the vehicle. In the early hours of 23 April, a malicious fire caused more than a million pounds worth of damage to the Munchie Snack factory in Dunmurry, despite the efforts of 50 firefighters with 9 appliances who attended to fight it. Due to the large quantity of flammable material and the open plan design of the building this was a difficult fire to deal with. Firefighters working in teams used breathing apparatus to enter the premises and fight the fire with jets of water. In places materials were stacked nine feet high and it took several hours to extinguish. A massive pall of thick, black smoke rose over the incident and a large part of the building, its contents and the roof were destroyed. On the same night at least three separate fires were started in different vacant flats in a large block in Derrycoole Way. Station Officer

---

290 *Belfast Telegraph.*
291 Nick Allaway.

Brian Greer, from Whitla Street station, said that about ten separate sections of the roof had been damaged:

> The vandals seemed to set them so that the flames spread up the stairwell, up stairs and into the roof void. We have been here a couple of times before with fire outbreaks but this is the worst damage so far. There was no one hurt in this fire but there is always the danger that the person who starts them will get trapped. That has happened before.[292]

**

Like a Stephen Spielberg Film

On 2 September residents in the Oldpark area of the city noticed a pronounced smell of gas. Almost immediately there was an explosion followed by a huge fireball, which damaged numerous houses. Sub Officer Jim Hamilton eventually traced the cause to a leak of a petroleum-based pollutant from business premises in Dunowen Gardens. The Annual Report of the Fire Authority described the incident:

> Oldpark, Belfast, 2nd of September 1991, 6.10pm. - Night duties for red watch started literally with a bang at 6.10pm on the second of September 1991. Over forty calls were received by control to fifteen houses on fire in a cul-de-sac off the Deerpark Road. As Westland's appliances were leaving their station firefighters could see the smoke rising from the incident, appliances from Central, Whitla, Springfield and Cadogan all formed part of the make up. Three jets, two hosereel jets, two no. 10 foam making branches, one mechanical foam generator no. 5 (MFG 5) and two breathing apparatus sets were used. The incident occurred when a suspected petroleum based pollutant caught fire in a river and travelled along the drains close to the houses. It was described by one householder, who ran to collect her children, as something from a Stephen Spielberg film. Large tongues of flame coming out of the ground and engulfing the houses. Thankfully, no casualties were reported, but for a while it was all hands to the pumps.[293]

---

[292] *Belfast Telegraph.*
[293] Annual Report for the Fire Authority for Northern Ireland

Station Officer Charles McAuley said:

> It was a very unusual fire in that it was spread over such a large area. The problem for us was not knowing the extent of the pollutant, where the source was and how big it was. There were garden fences and even houses partly on fire. The fire spread so quickly people couldn't see where it started and there was a lot of screaming because no one knew what was happening.[294]

At about 2.30am on 4 September, two furniture shops on the Shore Road were virtually destroyed by fires that were started by several incendiary devices. Firefighters with 15 appliances fought the fires, which began about 15 minutes apart. While there was little they could do to save the premises, firefighters did manage to stop either fire from spreading to adjoining buildings. In the early hours of 7 September Noblett's Wholesale Paint Warehouse at the junction of Carew Street and Witham Street was virtually destroyed by a major fire.

> I was on out duty at Knock and I was driving the HP and we got Kenneth Noblett's store over in Witham Street. And just how that fire had spread. I ended up using the cutters and cutting roller shutters and things so we could get into other premises and stop it spreading.[295]

Early on 13 September, a four-storey building in Wellington Place was badly damaged by fire. Firefighters with four appliances fought the fire for over four hours but despite their efforts three floors of the building were badly damaged and the roof caved in. On the morning of 18 September, Texas Homecare at Upper Galwally was badly damaged when six incendiary devices detonated. Firefighters spent an hour bringing the fire under control. Divisional Commander Jack Fell said that seven fire appliances were sent to deal with the incident, and six firefighters who went into the store had 'a very lucky escape' when another device exploded. They were 'very close to where it exploded and were lucky not to be injured. That's one of the risks of the job because it's hard to know if a fire has been started deliberately or as the result of an accident and we always hope this kind of thing won't happen'.[296]

---

[294] *Belfast Telegraph.*

[295] Joe Sloan.

[296] *Belfast Telegraph.*

Early on 10 October two incendiary devices detonated in Ross's Auction House, which had been operating in Montgomery Street since 1934. Firefighters confined the damage to the ground floor when the first device detonated at around 4.00am. Approximately three hours later a second device detonated, damaging both the ground and first floors. Twenty firefighters with four appliances took several hours to prevent the fire from spreading through the 150-year-old building. However, the ground and first floors were badly damaged; a number of antiques and some valuable furniture were destroyed. Divisional Officer Alex Withers said, had his men not got there in time, 'we could have had a major fire on our hands. There was a lot of damage but we prevented it from spreading further. The structure of the building is still sound'.[297]

Overnight on 11 October, there were several incendiary attacks on premises in the city. One of the worst hit was Leisureworld in Queen Street where four devices detonated. Firefighters with five appliances used jets of water to fight the fire. However, the contents of the first floor of the 25,000 square foot premises were virtually destroyed, causing about £500,000 worth of damage. A fire, caused by incendiaries, also damaged Donaldson and Little's in Great Victoria Street. City Reptiles in Hope Street nearby lost its electricity supply due to the attack, and several snakes and tarantula spiders escaped from their wooden and glass tanks when the water sprinklers caused the tanks to expand. The owner of the shop said:

> I have two nine foot pythons, about 40 or 50 other snakes and 50 Mexican tarantulas. Some of them have got out of their boxes but they are still in the shop, which is locked.[298]

Eventually all of the reptiles were returned to their homes.

Bomb in a Hospital

On 2 November a bomb detonated at the Musgrave Park Hospital killing two soldiers. The hospital was largely a civilian establishment but it also contained a military wing, guarded by an adjacent army base. The device exploded in a tunnel leading from the basement of the civilian section to

---

[297] *Belfast Telegraph.*
[298] *Belfast Telegraph.*

the recreation room of the hospital's military wing. It reduced the two-storey building above it to rubble and injured ten people, including the five-year-old daughter of a soldier. It was a Saturday afternoon and the dead and injured had all been watching the England verses Australia Rugby World Cup final. The bombers had sawn through the metal draw bars of a steel security door and planted the device, with a timer delay of up to 60 minutes, against the tunnel wall 20 feet away from the wooden doors to the recreation room's bar. Both soldiers were killed instantly by the device, which contained about ten lb of Semtex explosive.[299] The Fire Authority's Annual Report said:

> A happy time for most ended when two army personnel were killed and a number of people injured, some seriously, when a terrorist bomb exploded adjacent to the basement recreation area of the military wing of the hospital, causing severe structural collapse and starting a fire. Cadogan, Central and Springfield crews were quickly in attendance and engaged three jets and 16 breathing apparatus sets in rescuing a number of trapped personnel and recovering the fatalities. Senior officers of the Brigade were in attendance.

Just before 10pm on 4 December a 1,000 lb. device, packed into a wheelie bin in a van, detonated in the city centre. An estimated several million pounds worth of damage was caused to approximately 30 buildings within a 200 yard radius, including the Europa Hotel and the Grand Opera House. A nine-foot wide, six-foot deep crater was blown into the road adjacent to the Europa, and huge chunks of masonry and metal littered the street. Up to 30 cars were destroyed and the mangled remains of some of them were blown, still burning, on top of each other.

1992

POLITICAL TALKS RESTARTED at Stormont during 1992 but little progress was made. The Troubles continued and once again there were many shooting, bombing and burning incidents throughout the year.

---

[299] McKittrick, Kelters, Feeney and Thornton (1999).

Blast Incendiary Devices

Over the years the blast incendiary became a favoured weapon of the bomber, and firefighters in Belfast became quite experienced in dealing with them, even though the design and construction was constantly being improved:

> They gradually got more sophisticated over the years. It used to be condoms and a wee drop of acid, then they moved into lighter fuel, and then the clock and the timer and the battery and the light bulb and it all started to become more sophisticated. In the end there was Semtex in some that we were having to deal with, that would have been in the latter part of the Troubles where you wouldn't have gone near it or it would have blown your leg off. [300]

Armed bombers would often plant such devices inside buildings where they could be really effective in damaging the premises:

> It was usually a charge connected to a five-gallon can or a one-gallon can of petrol, and if it went off successfully it started a fire, a fierce fire. [301]

But another favoured tactic was to hang the device from the security grille over a window:

> Like a butcher's hook hung on the grille, with a canister of some kind of accelerant and a blast device on the outside, which explodes and blows the whole lot into the shop, so whenever we went, the fire was pretty well established.[302]

On 1 January there were several fires caused by incendiary devices, both in the city and in Newtownards, where firefighters from Belfast attended and supported crews from the local area in dealing with the incidents. Devices detonated in Top Man in Donegall Place and started a fierce fire. Firefighters wearing breathing apparatus entered the building with jets of water to fight the fire at close range. In all 30 firefighters with five pumping appliances, a turntable ladder and a hydraulic platform managed to control and eventually extinguish the fire. The premises were badly damaged, but fire spread to

---

[300] Colin Lammy.

[301] Colin Lammy.

[302] Gordon Latimer.

adjoining buildings was prevented. A second device detonated in Budget DIY on the Boucher Road damaging a number of rugs in the premises but the fire was quickly extinguished. In Newtownards the Suits Direct shop on the Strangford Road was virtually destroyed by fire, 11 cars were badly damaged at Edwin Long Car Sales in the adjoining premises and Martin Philips Carpet Warehouse was also damaged by fire. In the early hours of 2 January, Textile World in Boucher Crescent (next door to Budget DIY) was badly damaged by a fire that was started by one or more incendiary devices. The premises were well alight when firefighters arrived with six appliances and the entire stock was destroyed. Divisional Officer Sam McIntyre said that it was difficult to get access to fight the fire due to the security shutters on the building:

> The damage was extensive. It was heavily smoke-logged and there was obviously water damage because of using jets and hoses.[303]

On the night of 3 January an accidental gas leak from a tank of liquid petroleum gas (LPG) exploded in a shop on the Holywood Road. Blast and fire damage was caused to the entire block of shops but no one was injured.

On the night of 10 January, an incendiary device started a fire that badly damaged the Next clothing shop in Donegall Place. The first and second floors of the premises were badly damaged, as windows shattered and the intense heat warped metal beams. It took 45 firefighters with eight appliances to bring the fire under control. The Divisional Commander, Jack Fell, said that the fire was deep seated and firefighters 'took severe punishment' trying to locate it:

> It would appear that the fire started on the first floor. The second and third floor ceilings have collapsed and the roof is out completely. It's a mess. Luckily we managed to stop it spreading across to other premises.[304]

Ship Firefighting

Belfast has traditionally been associated with its docks and shipbuilding expertise. For this reason the city's firefighters have always been required

---

[303] *Belfast Telegraph.*
[304] *Belfast Telegraph.*

to deal with fires on board ship, both in the docks and at sea. One of the earliest recorded examples of this type of firefighting occurred on 25 December 1892. The 1,600 ton Minister of Marine loaded with 2,400 tons of coal had been towed into the harbour on fire. Superintendant Parker (the person in charge of the Fire Brigade at that time) reported in the fire attendance log book that:

> On our arrival I found that the whole of the cargo on the ship forward of the mainmast was on fire.[305]

On 4 March 1992 at 11.08am, fire control received a call to a fire in number five hold of the Crimson Star, an oil tanker that was in the Harland and Wolff Ship Building Dock at the time. On arrival firefighters saw that a well-developed fire involving residual oil in the bilges had resulted in thick smoke being emitted from the ship. A firefighting team, wearing breathing apparatus, was committed to climb 90 feet down ladders in order to attack the seat of fire, which was in the hold. At the same time an additional breathing apparatus team entered the adjoining hold with a water jet to cool the metal wall plates and prevent them from buckling. As operations continued, it was discovered that the fire had entered the cavity at the top of the double-bottomed tanks in the hold and a decision was made to use a foam-making branch in order to fully extinguish the fire. In all seven pumps, a foam tender, an emergency tender and a hydraulic platform attended, and a fire tug was mobilised to the area as a precautionary measure. This type of fire is particularly difficult to deal with because of the limited access, and the heat generated. Whiltla Street Station, due to their proximity to the docks, specialised in ship firefighting, although all firefighters in the city could be used to deal with a fire on board a ship if necessary:

> We got a few, because we were in Whitla, our area was the docks and there was some ship fires, which were frightening. There was one, myself and Paddy Hanna, we were on breathing apparatus and the fire was right down in the bottom of the engine room. You were right over the top of the fire going down, you couldn't put your hand on the rail because it would have just lifted the skin off you. So we went down, we had to go down the stairs with an MFG 5 (a foam making branch). The boiler had snuffed out and the oil was still pouring out, you know into like the bund tray

305 Macbeth (1954).

and it was sort of filled up, and then the thing ignited again and set fire to the oil in the bund tray. It was just black; you couldn't see a thing in front of you. The fire itself wasn't that big but it was just so hot. You had the breathing apparatus, the mask used to hang on a plastic thing, well that all melted. The whole thing melted, it was stuck onto my tunic and my ears were absolutely roasted, they are still scarred. In the sun they still bleed.[306]

** **

Late on the evening of 8 May a fire started in a house in Orby Parade. An eight-year-old girl, Lynne Kirker, woke up her 75-year-old grandfather and helped him out of the house. She then went back in and called the Fire Brigade, who attended and extinguished the fire. Neither the girl nor her grandfather were injured. Leslie Johnston was in charge of the incident and he said:

> It was only later that we found out what Lynne had done, going back into the house and bringing her grandfather out. She got away with it because she was so tiny. Because she was fairly low down, the flames and fumes weren't effecting her so much. Mr. Long had a scorched head and face, but we believe if she hadn't taken the action she did, he would have perished.[307]

Station Officer Kenny Leatham said:

> She had learnt what to do from the television ads – in the case of fire, get out, call the Fire Brigade out and stay out. She may have got them in the wrong order, but she is a very brave girl. We don't recommend that people take the kind of action that she did, however.[308]

Some time later Deputy Chief Fire Officer Stephen Walker, at Brigade Headquarters in Lisburn, awarded Lynne a scroll and certificate in recognition of her actions.

On 25 May, a memorial to firefighters who had lost their lives attempting to save others was unveiled at St Anne's Cathedral in Belfast. The Chief,

---

306 Dessie McCullough.
307 *Belfast Telegraph*.
308 *Belfast Telegraph*.

Ken McNeil, attended the ceremony with a number of firefighters, including Ken Spence and myself. Archbishop Robert Eames who unveiled the memorial said:

> The devotion to duty and the sheer professionalism of the Northern Ireland Fire Service has stood as an undeniable beacon of decency, service and support to the people of this province. Irrespective of class or creed, irrespective of political or religious loyalties, irrespective of location, the call to save lives and to protect properties has been responded to.[309]

The MV *Veronica*

During the afternoon of 8 June a fire started in a factory trawler, the MV *Veronica* at Harland and Wolff Shipyard. It took 19 hours to bring it under control. A fire service spokesperson said:

> The fire was in the lower hold of the trawler factory ship *Veronica* and 11 fire engines and 50 firefighters were sent to the scene. Ten water jets were used to cool the ship, five of which were on the fire tug on the seaward side.[310]

On 9 June, the trawler was towed, still on fire, from the Harland and Wolff deep-water dock to moorings in the Abercorn Basin where the fire was finally extinguished. Five appliances remained at the scene for several days, pumping out the boat, damping down and ensuring that the fire did not reignite:

> That was a bit scary. Charlie Hughes and I ended up on BAs, crawling up a corridor, we were crawling up this on our bellies and it was roasting hot, really, really hot and we looked ahead and I could see this red glow. It turned out it was the bulkhead at the far end, glowing red with the fire on the other side of it. And then every time we put a jet on, and hit it, the steam would come flying back towards us and we would have to, with our heads on the ground, let the steam go over our heads.[311]

---

[309] *Belfast Telegraph.*

[310] *Belfast Telegraph.*

[311] Stanley Spray.

There was a BA team in and they said they need somebody to get on the tugboat. Well, I was able to operate the tugboat monitor so I had to go and get on the tugboat. They said, just boundary-cool the side. But there was a squad inside and it was scalding. Now I didn't know but I was putting the steam inside on the boys. And then later on, we were back at it. So they needed a BA team to go down this ladder, at a jaunty angle, into this fish hole to look for water. They said they were pumping water on to this and they didn't know where it was going. The water came up to your waist.[312]

Fire Control

At that time the policy was for an officer based at headquarters to attend the control room during the larger incidents, to help with policy, operational interpretation and to support control operators in their decision-making. That was my role as the MV *Veronica* incident developed. However, I had been in the control room previously when it was incredibly busy and I knew the operators would be calm and professional, doing what needed to be done; and they were. So I did my best to make the tea and not get in the way:

> I could see the work they were doing and that was just gobsmacking. How the operators could sit there in front of the screens, hold a conversation with me over a problem that I needed to talk through with them and just operate everything, it was another world. They were extremely gifted professional people.[313]

> You thrive on being busy, you would rather have the busy moments because all the training comes out and all your teamwork comes out. If it got busy people were prepared to knuckle down and the teamwork was the thing that brought you through it. We all knew what was demanded of us and we all knew what we were meant to do. And yes there was a buzz. There was that air of calm but a controlled buzz. I've never experienced any time when we had to say to the public that we can't attend or that we won't attend. We never had that position; we were always able to juggle things about and, instead of sending four sending three so we

---

[312] Kenneth McLaughlin.
[313] Gordon Galbraith.

can send one somewhere else. And again it was the teamwork with the people on the stations, who we knew; you knew that a certain officer would release a machine for you right away. Or you knew of a certain officer who you would have to prompt. So there were all those intricate things, which were never said but were understood. And the control operators would be very focused. People wouldn't take breaks; they wouldn't leave their seat. Because at that seat they were doing a certain job, they were controlling Belfast or they were doing radios and they would get very focused and it would be very hard to shift them to take a break. So people got very focused on what was going on and trying to pass it on to somebody is nearly impossible, because it's all in your mind.[314]

Fire control can be seen by the general public as a call handling facility, but in reality it is much more complex than that:

The call will come in, you will ask the caller have they got a problem, you ask them for their address, look up the address while you are still talking to them, extracting more information, so you are sort of multi-tasking. You put the address into the system, you verify the address, and you ask them to repeat what the problem is. They'll give you an indication whether it's someone trapped in a car accident, a bin on fire or a house on fire. We will then make a decision of what machine is going to go based on location, how many are going to go based on the type of incident, and again the firefighters will be mobilised or sent out immediately. We will then track their progress, they will obviously give us information, telling us they have gone to the call, they have arrived at the call, do they require further assistance? If they don't they will put a message back saying they have dealt with the fire and they are now available to deal with other fires and then basically they will return to the fire station. And, if it is a part-time station we take details of who went out for payment, and how many went out and so on. In between times there will be other things to do, you'll inform the police, you may inform the ambulance, you'll inform local councils about different bits and pieces. So a whole variety of tasks that have to be performed. In addition to that you'll have to send supervisory officers to different types of fires. If it's

---

314 Jim Quinn.

a fire with, for example persons trapped, there is a certain level of officer must go out to give their experience and to control the whole incident. You have the ultimate discretion, which is one of the interesting things about the job, in that whilst the firefighters are on the ground or actually there performing the task, they can only perform the task if you give them the right number of people and the right equipment. So yes, you do have a lot of discretion and a good system, people are encouraged to use that.[315]

Control operators can also give the public life saving advice:

If we feel that the caller can't get out of the house, or they are trapped in the premises, or they are displaying to us over the phone some sort of issues about actually trying to get out of the house, we will give them basic fire safety advice. If they tell us for example they are in an upstairs bedroom and they are trapped, you might say to them, 'if you have access to a wet towel put it on the floor to stop the smoke coming through, put some form of indicator out the window so the fire service when they come know what location you are in the house'. We'll ask them for other details, we'll ask them can they safely make it out of the house and if they can we'll give them advice to get out and shut the door behind them and stay out. Or they've got children and they can't reach the children, they may tell us there's people trapped in different rooms. All sorts of different call handling advice, as we call it.[316]

And during the Troubles 'control were the 'unsung heroes'.[317]

By this stage of the Troubles, bonfires on 11 July were becoming bigger and more dangerous to life and property. In an atmosphere of rivalry and competition, both with 'the other side' and with other bonfire builders, it became commonplace for bonfires to be guarded overnight, and sometimes young people would even sleep in makeshift huts or gaps within the bonfire itself. Becoming increasingly worried about the situation and in the build up to 11 July the Brigade issued safety guidance for bonfires and Station Officer

---

[315] Jim Quinn.

[316] Jim Quinn.

[317] Roger Dawson.

Kenny Leatham said that the Fire Brigade is 'concerned that young people may be sleeping in makeshift huts or in bonfires while they are being built... obviously this could have disastrous consequences if the bonfire is lit while they are still inside'.[318]

Just after midnight on 10 July police were called to deal with a disturbance at a bonfire in the Ballybeen Estate. Fighting broke out and quickly escalated as young people guarding bonfires all over the estate joined in. A Transit Van was set on fire and the Brigade was called. Bricks and petrol bombs were thrown as the van was extinguished by firefighters and the police restored order. However the next few days passed without any serious incidents, even though the Brigade was as busy as always, dealing with the issues caused at this time of year.

The Freedom of the City of Belfast

On the evening of 21 August the Brigade was awarded the freedom of the city for their long history of service to the people of Belfast. The resolution of the City Council stated:

> Freedom of the city – The Northern Ireland Fire Brigade. Moved by Alderman H Smith, seconded by Alderman F Proctor and resolved – That the Northern Ireland Fire Brigade be and is hearby elected as a freeman of the city of Belfast in recognition of the contribution it has made to the life of the city and as a mark of the gratitude felt by its citizens to the personnel of the Northern Ireland Fire Brigade, many of whom daily risk their lives in the service of others.

More than 100 uniformed members of the Brigade took part in a formal parade in front of the City Hall before a special conferment ceremony in the council chamber. Councillors in full ceremonial regalia held a special meeting of the council at which the Lord Mayor presented the certificate of freedom in a silver casket to Chief Fire Officer Ken McNeil. At the civic dinner, following the ceremony, the Chief presented a specially commissioned Tyrone Crystal gift to the city from the Fire Brigade. Bob Pollock was the Station Commander of Central Station around that time and he had an involvement in persuading the council to make the award.

---

[318] *Belfast Telegraph.*

I can remember a lot of preparations for a parade around the city and a civic reception, so it was a big occasion. There were the usual jokes about well what does it actually mean, you can run your sheep through the middle of the city or whatever but there was a real sense of pride that we were being recognised for having done a good job well.[319]

It was I think an acknowledgement of all the service had done and it was well received. We rehearsed in the docks for the march past. There was a parade and the salute was taken by the Chief Fire Officer, the Lord Mayor and other civic dignitaries.[320]

However, not all firefighters appreciated the honour:

I remember it happening, yeah. I didn't even know what it entailed but we were given it, I remember that. Nobody really thought a lot of it, it was nice to have it but we didn't really know what it entailed. I didn't to tell you the truth.[321]

There was a certain amount of cynicism:

Obviously it was nice to get the recognition for all the stuff that you had gone through, that the council and the people of the city of Belfast, a way of saying thank you, and it was nice. But it wasn't putting any extra money in your pocket, which was more important in those days you know.[322]

And by the end of the night some of the people there had taken full advantage of the free bar:

He was drinking brandy and port, because 'he had a bad throat from all the shouting', and at the end of the night we ceremonially carried him out and put him in the back of a taxi and sent him home.[323]

---

[319] Brian McLintock.

[320] Chris Kerr.

[321] William McClay.

[322] Stanley Spray.

[323] Bob Pollock.

I had some sadness about that because I saw some people walking round the City Hall in no fit state in their uniforms. There was quite a lot of drink taken.[324]

I had the honour of attending the proceedings and helping one or two firefighters into taxis. No damage was done, everyone got home safely and, all things taken into consideration, the event seemed to be a real recognition of what the fire service contributed to the city during the difficult days of the Troubles.

The End of an Era as Central Closes

On 27 November 1992 Central Fire Station in Chichester Street closed. Built in 1894 this station was originally the headquarters of the old Belfast Fire Brigade and had been the location where generations of Belfast firefighters, including myself, had been trained. As the Brigade's Projects Officer I had a role in negotiating the move to the new purpose built station and divisional headquarters in Bankmore Street, as part of the reorganisation of the previous year. The facility would direct the new Belfast division, it would provide modern capabilities near to the city centre and it would take firefighters away from the courts complex, making them less likely to be caught up in attacks on those buildings. However, I did feel an emotional attachment to the old station, where I had started my career and was posted as a firefighter for the first four years of my service. For this reason I felt a certain kind of sadness when the station closed and I know I wasn't the only one. As the four appliances left the old Central for the last time they put on their two-tone horns and blue lights for the short drive round to the new station in Bankmore Street.

**

On 1 December a blast incendiary detonated in the stairway of Standard Supply in Ann Street, just minutes after a telephone warning had been given. Twenty-five people were injured, two of them seriously, and a fire was started. Four appliances from the new Central Fire Station in Bankmore Street were quickly on the scene, and 20 people were rescued from the first floor of the building by firefighters using extension ladders, as the fire was extinguished.

---

[324] Louis Jones.

The bombing and burning continued to the end of the year and overnight on 10 December two warehouses in Duncrue Street were badly damaged by incendiary devices. More than half a million pounds worth of damage was caused before firefighters could extinguish the resulting fires. Six pumps, a turntable ladder and a hydraulic platform were used to fight the fire. Sub Officer Kevin McCamley said:

> We had a certain amount of difficulty gaining access because of the roller shutters but once we got inside we were able to bring the blaze under control fairly quickly. The boys worked hard in difficult conditions.[325]

The following night a third fire was started in the same industrial estate in Duncrue Street, and another warehouse was extensively damaged. It took 40 firefighters with eight appliances more than 90 minutes to extinguish the fire, which was well alight when they arrived.

1993

MOVEMENT TOWARDS PEACE began to take shape in 1993 and there were regular meetings between the various parties. However, the republican and loyalist killings continued. In March the census showed that most people in Northern Ireland were segregated along religious lines and most of the population were living in areas that were more than 90 per cent Protestant or 90 per cent Catholic. In April a new £300,000 training facility opened in Westland Fire Station, ending the practice of having to use temporary facilities in different parts of the city, which followed the bombing that devastated Central three years earlier.

The New Central Fire Station and Divisional Headquarters
is Officially Opened

On 10 May the new Central Fire Station and A Division Headquarters based in Bankmore Street was officially opened. One of the highlights in the history of the Northern Ireland Fire Brigade, the project was the largest capital scheme undertaken by the authority to that date. The Chair of the authority, Mr D. Connolly OBE said:

---

[325] *Belfast Telegraph.*

The completion of the long awaited Central Fire Station and A Division Headquarters based at Bankmore Street, Belfast was finally accomplished in November 1992. Much excitement, tinged with a little sadness at leaving the historic building in Chichester Street accompanied the final departure, for the former Fire Station had served the citizens of Belfast well for 100 years. However, the new complex, housing all the essential departments necessary for the efficient operation of a modern firefighting unit, should prove even more effective well into the 21st century.[326]

The Secretary of State for Northern Ireland, Sir Patrick Mayhew, declared the new building officially open by unveiling a plaque and ringing the bell in the station's appliance room. Jack Fell was the Divisional Commander and, on his retirement on 1 June, Mick Malone succeeded him. Bob Pollock was the Station Commander. The division comprised six wholetime stations and one wholetime station with retained back-up, a complement of 14 pumping appliances and five specialist appliances, crewed by 400 wholetime and 20 retained personnel.

On 20 May the Grand Opera House was severely damaged when a 1,000 lb. explosive device, hidden under scrap metal in a skip lorry, detonated. Thirteen people were injured and damage was caused to many other buildings in the area. At the time it was estimated that the damage would cost up to £6,500,000 to repair.

Appliances Hijacked at Gunpoint

Throughout the Troubles, firefighters would be involved in incidents where shooting took place:

> We got a call one night to Albert Street and we were down at the Durham Street corner and the shooting started. Gordon and I jumped under the machine and the rest of the boys ran. We lay under the machine for about an hour.[327]

> One time we were at the Short Strand and they were shooting at us, and the thing I always remember about being shot at was

---

[326] Leaflet produced by the Fire Authority for Northern Ireland.
[327] Jimmy Armstrong.

you could never tell where it was coming from, and that always freaked me out.[328]

Every time I go through Central Station I remember lying on the railway lines with the bullets pinging on the platform above us. Whether they were shooting at us, or who they were shooting at, I don't know. It was at an incident and I just remember laying on the railway tracks hoping that there was no train coming.[329]

Sometimes there would be shooting at or near a fire station, particularly Ardoyne:

I've got one of the bullets out of the door. The front doors were all peppered. You used to be lying in bed at night and you would hear the rata tat tat.[330]

However, even in these, the most difficult of circumstances, the Brigade would always attend and attempt to deal with incidents:

You would never think of not going. You always thought of going and even when you saw somebody standing with a gun, instead of turning around and driving away you drove up and you tried to talk to them to see if they would let you get in.[331]

We were going to a call in an army base that had gone on fire, a hut. Two appliances and we approached the brow of a hill where we had to turn round a roundabout and back down again, and I was driving the second fire appliance. The first one was through before these two guys realised what was happening. But they were out on the road in front of me, with guns in their hands, balaclavas on their heads, obviously wanting us to stop. So, basically I threw the appliance down a gear, hit the accelerator, and the two guys jumped quite quickly out of the way and we got to the scene of the fire.[332]

---

[328] Louis Jones.

[329] Harry Welsh.

[330] Kenneth McLaughlin.

[331] Joe Sloan.

[332] Jim Hughes.

Quite often you would be threatened with a gun and told to let a particular fire burn:

> We got called out to this single decker bus on fire across the Falls Road, just at the Falls Road Library, so really we were getting busy with the hoses and the usual stuff. Then this guy appeared out of the shadows and he said to us 'Provisional IRA, fuck off!' And he motioned that he had a gun in his pocket and we said 'OK mate we are pulling out'.[333]

> We turned up to a particular incident and there was a funny uneasiness in the air, there weren't any kids, which was unusual, but there was something funny going on. It was a house fire and we were approached by a couple of gentlemen. They came down and it was a question of, hands in the pockets and, I have a weapon here and we want you to leave. We don't want this incident put out.[334]

And sometimes an appliance would be taken from you:

> They took our fire engines a couple of times on the New Lodge, put a gun to the officer-in-charge's head telling him they wanted the fire engine, we just used to give them the fire engine.[335]

One of the worst examples of this happened in 1993 when the funeral of a UVF member, on Friday 2 July, was the catalyst for 15 hours of rioting in the city. There were numerous petrol bomb attacks and at least 20 attempts to kill police officers in gun and grenade attacks. More than 70 vehicles were hijacked and burnt, and the rioters attacked firefighters who were responding to fight the fires. During the night an appliance from Cadogan Fire Station was hijacked at gunpoint and burnt on the Donegall Road. The annual report of the Fire Authority described the incident:

> On 2 July 1993 at 11.39pm, the Brigade received a call to a car on fire on the Donegall Road, Belfast. On arrival the fire crew was confronted by a hostile crowd, burning cars and barricades obstructing the road. Armed and masked men surrounded the

---

[333] Nick Allaway.

[334] Gordon Galbraith.

[335] Louis Jones.

fire appliance and held the officer-in-charge at gunpoint. After driving the appliance along the road for a short distance, the officer-in-charge was released. The masked men then blocked the road with the fire appliance and set it on fire. The appliance and its contents were completely destroyed.

Divisional Officer Alec Withers, in condemning the attack the next morning said:

> One of the crew members had a gun put to his head and then fired a short distance from his ear. He was then forced to march down the road with his hands in the air. The other crew members were then rounded up and asked to drive the appliance across the road before being ordered to leave the area. A number of crews were stoned while fighting fires and this morning we found several appliances had broken windows as a result.[336]

Speaking on behalf of the FBU, Executive Committee Member, Archie Culbert appealed to the community to give firefighters safe passage through the troubled areas, emphasising that lives could be at risk:

> The saving of lives from the dangers of fire and the protection of property and possibly jobs depend on the immediate response of fire appliances and their crews. Any delay caused by firefighters having to seek safe passage could have serious if not fatal consequences. The union calls on the whole community to reflect seriously on the implications of last night's actions.[337]

The firefighters concerned were treated for shock and, in an early example of this kind of intervention, were offered counselling through a post incident debrief. A weekend of violence followed with 35 shooting incidents, 11 petrol bomb attacks and 63 vehicles being hijacked. Dozens of vehicles were burnt and several premises were damaged by fire.

Frizzell's Fish Shop

At about 1pm on 23 October two IRA members, wearing white coats to give the impression that they were deliverymen, carried a brown box measuring

---

[336] *Belfast Telegraph.*
[337] *Belfast Telegraph.*

approximately 15 inches by 18 inches into Frizzell's fishmongers shop on the Shankill Road. The shop was full of people on a busy Saturday afternoon. Without saying a word they left the box, which contained an explosive device, on the counter and turned to leave. The device detonated almost immediately killing ten people, including one of the men who had planted it. Two children and four women were among the dead and in all 57 people were injured, many of them seriously. Among the injured were a 79-year-old woman and two male toddlers. John Frizzell, the owner of the shop, was killed and his daughter, who was in the shop at the time, died with him. However, his wife who had left the shop a few minutes before the explosion survived. The IRA said later that the targets for the bomb were loyalists meeting in an office above the shop. The three-storey terraced property and parts of the adjoining premises collapsed into a huge pile of debris, throwing dust, dirt and rubble across the road and into the air, bringing heavy masonry crashing down and trapping a number of people inside. Rescue teams, made up of Brigade personnel, police and civilians were organised under the direction of senior fire officers and five hours were spent in the delicate rescue operation, utilising chains of local people to move the rubble.[338] Ambulances carried the injured to hospital. Firefighters used heavy lifting equipment and their bare hands in the attempt to rescue the injured, and a digger driver, Winston Armstrong, who was working on a nearby building site was drafted in to help. Mick Malone, who was at the incident said:

> He showed a lot of expertise in handling the machine in very difficult circumstances. The whole building had collapsed with the exception of the roof and part of the chimney. We gave Mr Armstrong directions but he had to disturb a lot of timber and pull it out. Pieces of wall were falling down around him. I could not praise him highly enough.[339]

From time to time firefighters would call for silence as they listened for signs of life but to no avail, and eventually the operation became one of body recovery as the bodies of ten men, women and children were carried from

---

[338] Annual Report for the Fire Authority for Northern Ireland.
[339] *Belfast Telegraph.*

the collapsed building.[340] One eyewitness said, 'It was absolute carnage. I saw a young boy with his head blown off.'[341]

> I was out on duty in Springfield, as a Sub in charge of the two machines. I heard a bit of a bang and then we were called to a 77 actuated on the Shankill Road. We drove over and it was mayhem! The bomb had just gone off and it was just people all wandering around, their faces all covered in blood. You can imagine the state they were in. I said to the boys just climb out and do what you can, we'll get more help. Then we realised the shop had collapsed and there was bound to be people under it and half of it was on the pavement and in we piled. We hauled and trailed and hauled and trailed, it was just bedlam. Then we started pulling out body after body, the old man was the first, he had no eye, he had lost his eye, he was just about in one piece; that was the owner I think. Then we started to find other people, some looked as if they had just gone to sleep, others were just blown to pieces, completely blown to pieces. Stretcher after stretcher, this went on for I don't know how long, we just kept finding more and more of them. There was a wee child that was just about on its way out, and one of the boys said that child has just died; we didn't get to it in time. They were all buried underneath all the rubble. Then eventually Mick Malone said 'there is a whole load of blood on that wall Billy, I think there is still somebody in there'. The boys dug and dug and then they found what was the bomber, right at the bottom, just a pulp. The whole thing came down like a deck of cards you see.[342]

The UDA/UVF then carried out a wave of revenge killings against Catholics in which eight people died. Among them was Robert McClay, known as Rory, who was shot on his doorstep at around 9.30pm on 7 December. One of the terrible ironies of the Troubles was the fact that he was the brother of Billy McClay, who was the officer-in-charge of the appliances from Springfield Road Fire Station, which formed the first Brigade attendance to Frizzell's, and who did so much to try to help the casualties there: 'my young brother was

---

[340] Annual Report for the Fire Authority for Northern Ireland.
[341] *Belfast Telegraph*.
[342] William McClay

murdered on his doorstep, probably as a result of the bombing. He opened his door and a guy with a gun, an automatic gun, just shot him dead'.[343]

It was thought at the time that this would be the end of the peace process, but despite these expectations, or perhaps because of the horror and futility of the explosion and its aftermath, the peace process continued.

1994

ALTHOUGH THERE WAS movement towards a ceasefire, in the first half of 1994 it was not a smooth process and the Troubles continued.

In the early hours of 1 January a series of ten incendiary attacks across greater Belfast resulted in millions of pounds worth of damage and put 200 jobs at risk. Twenty-eight appliances were required to deal with the fires and additional crews were called into the city from Ballyclare, Holywood, Bangor, Lisburn and Glengormley. The fire service was stretched in dealing with the sheer number of incidents over such a short time. The morning's work started at 12.45am when a call was received to a fire at B&Q on the Shore Road. The premises were well alight when firefighters arrived and they fought the fire with jets of water, including one from a turntable ladder. Despite their efforts the building was virtually destroyed and part of the roof collapsed. At 12.52am a fire badly damaged Etam and Tammy Girl in Donegall Place. Smoke and flames could be seen coming from the skylights and neighbouring premises were damaged by smoke. A fire, which badly damaged the interior of the premises and its contents, was started in Primark at Castle Junction at 1.25am. At 1.40am a fire virtually destroyed Crazy Prices in Dunmurry:

> They targeted Crazy Prices and it was like a massive fire, and you could see the signs of collapse, so they pulled us out of it. They had sent us in with breathing apparatus to tackle the fire because it was going rightly. The buildings just went up very quickly. I think it was ones that they planted inside the shop itself; they generally planted them in amongst paint tins or combustible materials.[344]

And, at 2.06am, Texas Homecare in Newtownabbey was badly damaged by fire. At 2.25am Gino's Clothes shop had its contents destroyed and at 2.38am a rack of clothes was destroyed by fire in the Issue Store in

---

343 William McClay.

344 Nick Allaway.

Castle Lane. At 3.37am, in an attack the motivation for which was hard to understand, a number of books were destroyed in the 200-year-old Linen Hall Library. Particularly ironic was the fact that an award winning collection on Northern Ireland's Troubles was housed there. The library was founded in 1788 and it contained thousands of books and pamphlets, some of which were irreplaceable. Two incendiary devices detonated in the library and between 800 to 1,000 books were destroyed. Thankfully the Brigade managed to get there quickly, despite being particularly busy, and the fire was extinguished before too much damage was caused. Even though the fire came perilously close to particularly valuable items, the books lost were of general interest and the award winning collection was undamaged. The Chief Librarian said: 'We owe a great debt to the Fire Brigade. If they had arrived 15 minutes later, the flames would have spread into a room where priceless publications are stored'.[345]

At 3.12am Junior's Store in Castle Street was damaged by fire and at 3.37am a fire, which was quickly brought under control by the Brigade, damaged Chicago in Ann Street. All of the fires were extinguished by breakfast time but firefighters continued to dampen down premises for a number of hours after that. Of the 80 firefighters that dealt with the various incidents one was taken to hospital suffering from smoke inhalation.

The bombing and burning continued, but on 31 August the IRA declared a ceasefire from midnight:

> In August 1994 the Provisional IRA declared 'a complete cessation of military operations' which, though it was a long way short of renouncing violence, was widely welcomed. It was followed a month later by a Loyalist cease-fire. The road to peace would be tortuous, involving many walkouts and public arguments. On the streets, extortion, kneecapping and occasional murders continued. But whereas the number of people killed in 1993 had been 84, the toll fell to 61 the following year and nine in 1995.[346]

In 1995 the number of deaths due to the Troubles was in single figures for the first time since 1968. In Belfast seven people died compared to 30 in 1994 (22 from January to July and eight after the ceasefire was called in

---

[345] *Belfast Telegraph.*
[346] Marr (2008).

August).[347] However the Troubles had not gone away, ceasefires came and went and the deaths, albeit at a much reduced level, continued, with both assassinations and revenge killings. In 1998, as the Good Friday Agreement brought a more stable peace to Northern Ireland, the real IRA killed 29 people in an explosion in Omagh. And even now, well into the 21st century, Troubles-related incidents still occasionally result in rioting, and from time to time an improvised explosive device (IED) is found, perhaps in a public area or attached under a police officer's car. But, for the fire service in Belfast, work started to normalise in 1994.

For me there was also a significant change in 1994, because after 25 years service in the Belfast and Northern Ireland Fire Brigades, I took up a new position in the Lothian and Borders Fire Brigade based in Edinburgh.

---

[347] McKittrick, Kelters, Feeney and Thornton (1999).

PART SIX

The Damage Done

Lasting Effects[348]

THE STRUCTURAL DAMAGE caused by explosions and fires in Belfast during the Troubles is plain to see. You just have to walk through the city centre to see the new buildings, which were built to replace the often Victorian older ones that were destroyed between 1969 and 1994. The financial cost ran into many millions of pounds; however, the death and physical injury to so many people of the city caused much more lasting damage. One thousand five hundred and forty people lost their lives in Belfast as a result of the Troubles[349] and those injured, many of them seriously, number thousands more. Even more insidiously, the often invisible psychological damage affects many thousands of people. The Troubles had a profound effect on the population of Northern Ireland. In an article on 23 March 1977 entitled 'The Troubles and Our Children', the *Belfast Telegraph* reported that many children accepted violence as a way to pursue political ends, six-year-olds could understand the realities of segregation and there had been a major increase in the number of referrals of children to Belfast's educational psychologists. In 2015 a review of the trans-generational impact of the Troubles on mental health in Northern Ireland estimated that about 213,000, out of a population of approximately 1.5 million adults, were badly affected by the years of conflict. Furthermore it concluded that nearly half of the cases of severe mental health issues in Northern Ireland were directly associated with the Troubles. Even worse, while almost 30 per cent of the population suffered mental health problems, almost half of which related to the Troubles, the legacy of the conflict was also connected to suicidal

---

[348] The quotes in this chapter are anonymous, in order to protect the privacy of the contributors.

[349] Sanders and Wood (2012).

behaviour, which was likely to be transmitted to further generations.[350] Belfast firefighters are not immune to this legacy.

What Ever You Say, Say Nothing

The firefighters of Belfast dealt with some of the worst atrocities people could inflict on other human beings for more than 25 years. They dealt with rioting, bombing, blast incendiary devices, petrol bombing and incendiary attacks, while at the same time delivering an essential emergency service to all of the communities in the city, without fear or favour. As they did this they could be kidnapped, stoned, petrol bombed, threatened or shot at, and there is little doubt that the things they had to deal with over a prolonged period of time had an impact on them. The following words are from these firefighters:

> It seemed to be constant pressure.

> The two appliances went to McGurk's bar on North Queen Street and I remember old hardened firefighters coming back from that and they were in tears. There was another explosion at NIE Headquarters, the electricity service headquarters at Danesford in which a couple of young females were very, very badly injured. Again the same people came back in a state of shock, in tears. Obviously in those days there was no recognition of stress or post-traumatic stress syndrome or anything of that nature and people were expected to just get on with it, without a break.

> I saw two fatalities, an elderly mother and her son, twenties. It was accidental, it was a cigarette. That morning they were emigrating and that's the tragedy of it, you see. That hit home for me.

> Actually I will not watch horror movies, I will not watch murder, I will not watch serious murder that's sadistic. I'll watch, with *****, something like *Midsummer Murder*, that's a cartoon, but I will not watch something which has a bearing on life. I don't want it jogging my mind.

> There were some fires that I still have wee sort of flashbacks about, you know. I mean I'm out 18 years, just some things that

---

[350] Ulster University (2015).

happen on television you just say, oh, that's very like; it just clicks back into your head again.

Just realising then, as you drove back to the station, nobody speaking or anything, and having a cup of tea, it didn't seem to bother me at the time but I always remember the next night the guy's photo was on the front page of the *Telegraph* and then I had a lot of sleepless nights for a week or so after that. And I've never forgotten that.

Sometimes an especially difficult incident, not necessarily connected to the Troubles, can be particularly traumatic for some individuals:

It was a Friday afternoon, I was a Station Officer in central and Jack Fell was the Area Commander. It was just coming towards the end of lunchtime, Jack phoned down and asked me to go up and see him. He said I've got a call from the police and they have a rape victim who's been murdered, but the coroner wants her removed from where she has been found.

What had happened was off Sandy Row, blocks of two-storey flats with a stone staircase to the top flat and then a boundary wall close by it. In the triangle formed by the stairs and the wall of the premises and the ground, there was a little room, but at the back between the stairs and the boundary wall you could squeeze in. So this girl and her assailant must have squeezed in. She had been murdered and then bound and gagged using her own tights. She was bound in them almost in a foetal position, with her arms and her legs up so they couldn't get her back out the way she had come in. So we were asked then to remove the railings you know at the front. We removed the railings, got the girl out and that was it. But that was difficult to deal with at the time because, I suppose, you don't have the adrenaline. I had real difficulty putting it into a box. Whether it's fighting fires or road traffic collisions or whatever else, and there have been other things, that have been maybe terrorist related, where there is a feeling of anger but you still deal with it and you move on. I just couldn't understand how one individual could do what had been done to this girl. After all of those years and different incidents that's the only one that I dwelt on.

And firefighters often have difficulties when children are involved:

> It's difficult to pacify and comfort an adult but actually you can't deal with children, they scream, they scream, that's all they do they scream.

> We got a call, early hours of the morning, to a two-up two-down house off Sandy Row. It was a persons reported, and whenever we found the woman upstairs, I found her upstairs, and she was lying face down on the ground and she was black from the smoke and she was obviously dead. I noticed a head underneath her again and in case there was any life there I turned her over and there was an almost perfect baby in her arms, also dead. So I have nightmares about that sometimes.

However, there was a strong cultural imperative to deal with the emotions that this type of emergency work inevitably raises, through a masculine view of strength.

> There was a thing at the time, and I think it's probably still true. To get upset about something like that would be a weakness, and you didn't show weakness. It wasn't really talked about. There was that much happening, there was that much other things going on. We were just so busy.

Some people were able to cope at the time:

> I think in the Fire Brigade you have to accept that stuff happens. At the same time it was unfortunate, the horror of it and the matter of factness of it really. You become, not with hindsight, but you become at the time immune to it.

> Because you were young, when you look back on it we joined when we were 18, so we were kids basically, thrust into this man's world at a very crazy time so you sort of have to grow up fairly quickly don't you?

> As a young 18-year-old it was still a big adventure, it was still good craic and good fun, getting out to calls. We were busy and everything was an adventure. I never can recall being frightened,

getting into situations where we were unsure but never being scared. Thinking we are sort of indestructible here, we can deal with anything that's thrown at us.

There was a lot of people, all crying, but I don't consider that was me being macho or whatever, it wasn't, it just, you know I was obviously touched by it. I was extremely sad but there really is a bit of, just, you have to move on you know, and you always feel guilty for not displaying more emotion.

He drove me up to my Mum's house and my Dad's house and we sat outside and we both cried in the car. Obviously emotions had just got the better of us and we cried in the car, the two of us. So that had a real big profound effect on me. It was pretty blunt and brutal. But it didn't do me any harm because after that I sort of felt I could, well I could probably face most things. The only other thing that really upset me were children. But adults and stuff didn't really affect me that much. I was able to go into road traffic accidents and put my hands on people, where other guys were a wee bit standoffish it didn't really bother me. (Either you get the ability to put it into a compartment somewhere or you can't cope with the job.) That's right, you would just leave.

But, for some, there was a cost:

The Hayloft Bar, one Saturday night, it was in High Street on the second or third floor. We had been to bombs before, they seemed to come one a day, two a day, sometimes three a day, and for me being young I felt that inside me was just getting tenser and tenser and I thought I can't cope with this, and little did I know that for us as firemen, and me particularly, it hadn't really started. It was only starting to grow.

I must admit that that was something that didn't really affect me for over 20 years but then it came back into my memory and even talking to you now I can feel myself getting a bit anxious about that. So I went to the doctor and he gave me stuff to get me through. You were able to cope with it then, there and then, and you think why is this annoying me now? Even incidents that weren't connected to the Troubles annoy me and I think back and I say to myself could I not have done something better? Could I not have done something more?

The one thing it did to me was to make me so hard that I had no sympathy.

And, some had real difficulty in coping:

Then I had to go on the sick, the stress sort of thing started firing away, you know.

Well, there was no such a thing as stress in those days, that word didn't come in, you didn't know about it, there was no counselling, you just got on with it, you went home and had the nightmares. Some people didn't cope. There were guys who joined the job and within two years they had left it again; they couldn't hack it. I don't really know how I dealt with it because sometimes it affected me. You know you went in like the big macho man, like that was all right, but it really wasn't all right. You had flashbacks and stuff.

You tended to bottle it up, but some of our guys saw appalling sights over the years and there is no way in the world that you can just compartmentalise those things and put them away forever. At some point that has to find an outlet.

The Use of Alcohol

One of the ways that firefighters dealt with the things they faced was through the use of alcohol:

People had to find their own ways of coping, and the stress that a lot of the guys were under, but wouldn't necessarily admit to, because it wasn't the manly thing to do. It had to find an outlet somewhere and I'm quite convinced in my own mind that the drink culture, such as it was, was very heavily related to that.

You went out for a beer with your mates.

And not always off station:

If I can go back to my first day. At 12pm everybody disappeared, this happened for the first two days, everybody disappeared.

I thought where the hell is everybody, have I missed a call? The machines are still sitting there in the engine room. It wasn't till the second day I heard a bit of laughter. I thought what's going on there? Beside the galley was a room and it was always kept locked and it turned out the guys had a bar in there, with a cooler.

We liked our food in those days, any excuse at all for a steak supper so, I'll not say every set of duties, but at least once a month we would find some excuse to have a proper sit down slap up feed, and one or two good chefs on the watch as well, good with a steak. But the cases of beer always appeared alongside the steaks, it was just par for the course. I really do think that was a coping mechanism and it probably had evolved out of the Troubles. And certainly for the guys who had been in the job a bit longer than us, that was part and parcel of what they had grown up with in the service. I am quite sure that it was a coping mechanism.

On my first day on duty as a firefighter on Red Watch Central, in 1971, I was walking past the pump in the engine room at about 9.00am when a voice said, 'would you like a drink.' I looked up and sitting on the roof of the appliance were three firefighters, at least one of whom would go on to hold a relatively high rank. Thinking it was some kind of initiation test I said, 'why not' and climbed up to join them. The water tank of the appliance was filled through a cap in the top of the vehicle in those days, and one of the firefighters screwed off the cap and there were a number of s hooks formed from wire coat hangers and hooked over the lip of the tank opening. Attached to each hook was a piece of string and one was pulled up. It had a bottle of beer tied to the end of it and this was quickly opened and offered to me. So there we were, the four of us sitting on the top of the fire appliance at 9am on a first day shift, drinking beer.

They gave you a cup of hot wine, drink that son, it'll do you good. If you didn't drink they were suspicious of you, they didn't trust you.

It's the only thing that kept them sane was smokes and a drink. **** ***** sent crates of beer in, Guinness and stuff, went over to the station. I was a Pioneer, wore my pin and all, until Bloody Friday. We were going home, it must have been about 9.00am. Of course you know what ****** was like, into the Ivanhoe. What do you want? A Coke, 'aye' he says, 'you'll get a Coke all right,' and we were in a bad old shape that day. That really annoyed us you know. The next thing ****** brings down the Coke and there was a glass of vodka in it. That was the first drink I ever had, I broke my pioneer pledge.

My mind was just a blur. So eventually we got back to the station and everybody was trying to pull themselves together, get away. I went straight down to a bar in the town and drank as much as I could drink, and I rang the house to say what had happened and I drank and drank and drank; it didn't do me a bit of good. I wanted to put it out of my head.

We all decided we would go anyway and get this out of our system, and we were in ****'s house for 8.00pm and we didn't leave until half four in the morning. We drank bottle after bottle of vodka and nobody was drunk, nobody was drunk. Because we got half way into it and we started to talk about it then, and we went through the whole thing. I think that was the big saviour with the Fire Brigade. But I remember that night sitting drinking and we were drinking vodka, unbelievable the amount we drank.

Horseplay and Humour

There was always a great deal of humour in the Brigade and it was inevitably a dark humour, particularly in discussions following difficult incidents or fatalities:

The old Fire Brigade in Belfast was very much a working class thing and these guys were just naturally funny, and they had this brutal way of looking at things and doing things.

I think that's where the black humour would come in, you know as a coping mechanism.

I always remember the messing about. There was always a lot of messing about at night, the water fights and the burning newspapers shoved under the door. We used to climb out the window and go along the ledge. I think it was letting off steam, you had to you know, some of us were quite crazy and you did things you wouldn't dream of now.

Firefighters were always playing practical jokes on each other, and sometimes their officers:

The first one I remember is one you had an involvement in. We were heading up the Grosvenor Road with Big ****** trying to get another man's coat on. We had changed tunics, Big ****** was about six foot one tall and ****** was about five foot nothing, and ******* couldn't understand how he couldn't get into the other coat. We were actually going to a bomb on the Grosvenor Road and I remember we passed the bomb, drove past it and turned and came back down again to the Sandy Row Bridge. And we got to the corner, probably about forty yards away when the bomb blew up, and it blew me and the rest of the guys up, and our heads hit the roof of the fire engine. No one hurt and the incident with ****** and his coat made it a bit of fun.

I remember one night in Central taking ******'s 50cc motor bike up to the top floor and at about 4am, ****** came into the bedroom on the bike with the headlights on and revving up the engine.

There were lots of things about practical jokers on fire stations. I suppose ****** springs to mind, the things he would have done, all simple things. I remember he put cling film over the two urinals in the downstairs toilets in Central, and went to the extent of taking the light bulb out at night time, and somebody would come in and flick the switch and you could hear them flicking the switch, and the light didn't come on but they would go over and have a piss in the urinal. All you heard was druuuum and their feet would get soaked. It was wee things like that, and putting foam concentrate in the cistern and waiting till somebody sat down and then reached across and pulling the chain so that the foam came up round them.

The TL was deployed and ****** asked for volunteers to tie chair knots. So of course all the young recruits, fresh out of the training centre, wanted to show their expertise with knots. So two of us tied knots round ourselves, chair knots, and we were hoisted in the TL, just to show that it could take the weight of it. It was only at that point that we saw the jets being rolled out as we were dangling in mid air, not being able to do anything about it. So needless to say that was another soaking.

Big ****** and the light switch, he used to come in, in the mornings, and crank the door open and switch the lights on and wake everybody up. And they took the cover off the switch; that was the last time he did that.

And the water fights had to be seen to be believed:

I mean the water fights that used to go on used to be unbelievable. Guys would be sacked now, left right and centre, if they done anything like that.

Any God's amount of water fights. I can still remember the watch being called down and given a bollocking in the engine room and the water was still running down the pole from the pole shaft. I suppose the bit is; it was great craic. It really was, when you were part of it. I was 18 when I joined the fire service, so you know for the first four or five years sure it was just great craic. But, bin bags full of water taken to the top of the tower and trying then to drop them on people coming out of the TV room. And you think, if that had actually hit somebody, it could kill you, you know. The weight, I mean two men are trying to struggle to get this bin bag up onto a windowsill to be able to drop it out. But luckily people had the sense not to tie the bin bag so that if it did hit you there would be some sort of release. Although this sounds stupid, you could actually hear it because it was coming from such a height, you could actually hear the plastic bits of the bin bag flapping. This thing was just awash with water, but yeah, stupid I know and childish, but great craic.

He called us all down to the engine room, and the water pouring out of us. We all stood there soaking, saying 'what water bombs? We don't know anything about that.

Mutual Support: Talking it Through

However, the main way of dealing with things was by talking it through with other members of your watch, comrades who understood, because they had been there. For some reason adrenaline seemed to make you hungry and I remember coming back from difficult incidents on many occasions, making the tea, toast and jam, sitting around the mess room and talking, often with a great deal of dark humour thrown in.

> You came back, you had a cup of tea, and you sat round and you talked about the incident until you got it out of your system. And, when you went home at 9.00am the next day, I always felt OK. Calm because you discussed it right away. You didn't let it mull in your mind so nothing festered. I think that was one of the good things; that you talked openly about any incident you went to and you always made a laugh of it. See if you didn't, I don't think you would have been back the next night. You couldn't have coped. It was very black humour.

> I suppose talking too, where there would be people who would say, 'yes I remember something like that, I remember an incident from the '70s, etc.' And there were a number of senior firefighters on the watch held in very high regard, and just that reinforcement that it wasn't unique, it was something that other people had experienced.

> Sitting with a cup of tea in hand, just talking things through, not in any formalised way. It was just the informal chat, often with a good bit of banter and ragging thrown in. That was a very important mechanism.

> I don't know how we survived it to be honest, some of the things we did. But the camaraderie in the job sort of made up for all the bad times. There were so many good times, I mean the guys were like sort of brothers to you, you know. You looked after them and they looked after you. That was what I loved about the job really. You looked after each other.

> It was the people, and maybe the circumstances, you were together with great guys, you were all in the same boat, doing the same thing. I mean we all had the same problems basically, and if you didn't look after each other, there was nobody else going to look

after you. They talk about a band of brothers but I suppose each watch was a wee band.

If you were at a bad incident you came back and everybody would sit down and talk. Well, black humour, and then you sat and talked about it and I think that got a lot of things out of a lot of people. They could see that everyone was affected, they weren't the only one that felt the way they felt, you know. And I think that was a big thing.

Most of us just got on with it, and coped with things at the time, although some found it more difficult than others, and some just left the job. There was a lot of drink about and certainly the horseplay and humour allowed you to let off steam, even though some things that went on weren't appropriate in today's terms, and certainly wouldn't be accepted in current workplaces. However, the most powerful coping mechanism for most firefighters, before professional counselling was brought in, was the ability to sit over a cup of tea and talk things through with your mates.

For me that was both a necessity and a privilege.

Bibliography

Allaway B., Culture, Identity and Change in the Fire and Rescue Service, The Institution of Fire Engineers, (2011).

Bailey V., The First National Strike, contained in Forged in Fire: The History of the Fire Brigades Union, Ed. Bailey V., Lawrence and Wishart, (1992).

Bew P. and Gillespie G., Northern Ireland: A Chronology of the Troubles 1968 – 1993, Gill and Macmillan, (1993).

Boyd A., Holy War in Belfast, Anvil Books, (1969).

Broadhurst W. and Welsh H., The Flaming Truth: A History of the Belfast Fire Brigade, Flaming Publications, (2001).

Broadhurst W. and Welsh H., Out of the Fire: A History of the Fire Brigade in Lisburn, Flaming Publications, (2004).

Elliot S. and Flackes W. D., Conflict in Northern Ireland, Blackstaff Press, (1999).

Englander D., The Fire Brigades Union and its Members, contained in Forged in Fire: The History of the Fire Brigades Union, Ed. Bailey V., Lawrence and Wishart, (1992).

Fay M. T., Morrisey M. and Smyth M., Northern Ireland's Troubles: The Human Costs, Pluto Press, (1999).

Gillespie G., Years of Darkness: The Troubles Remembered, Gill and Macmillan, (2008).

Gillespie G., A Short History of the Troubles, Gill and Macmillan, (2010).

Hennessey T., The Evolution of The Troubles 1970–72, Irish Academic Press, (2007).

Hepburn A. C., A Past Apart: Studies in the History of Catholic Belfast, the Ulster Historical Foundation, (1996).

Keane F., View from the Outside: A Troubles Archive Essay: The Arts Council of Northern Ireland: (2011).

Keenan B, I'll Tell Me Ma, Jonathan Cape, (2009).

Macbeth H. The story of the Belfast Fire Brigade, (typescript held in the Ulster Museum), (1954).

Maguire W., Belfast; A History, Carnegie Publishing Ltd, (2009).

Marr A., A History of Modern Britain, Pan Macmillan, (2008).

McKittrick D., Kelters S., Feeney B. and Thornton C., Lost Lives, the Stories of the Men, Women and Children who died as a result of the Northern Ireland Troubles, Mainstream Publishing, (1999).

Parkinson A. F., 1972 And The Ulster Troubles, Four Courts Press, (2010).

Sanders A. and Wood I., Times of Troubles: Britain's War in Northern Ireland, Edinburgh University Press, (2012).

Smith S., 3-2-1 Bomb Gone: Fighting Terrorist Bombers in Northern Ireland, Sutton Publishing Ltd, (2006).

Styles G. and Perrin B., Bombs Have No Pity, William Luscombe Publisher Ltd., (1975).

Wharton. P., First Light: The Explosive Memoirs of a British Army ATO, The Book Guild Ltd, (2012).

Wright A., Burning Issues: A Belfast Fire Officers Memoirs of the City Firefighters During 'The Troubles' 68–88, Rosepark Publishing, (1999).

Reports

Boyce H.R.C. The Maysfield Leisure Centre: Report and Investigation into the circumstances leading up to the fire on the 14 January 1984, Department of Economic Development (Northern Ireland), (1984).

Towards A Better Future: The Trans-generational Impact of the Troubles on Mental Health: Prepared for the Commission for Victims and Survivors by Ulster University (March 2015).

Newspapers

Belfast Telegraph.

The Irish News.

News Letter.

The Irish Times – from May 1 1980 to May 14 1980 (Northern Ireland newspapers not printed due to industrial action).

The Scotsman – from May 1 1980 to May 14 1980 (Northern Ireland newspapers not printed due to industrial action).

Television Programmes

Forgotten Heroes: Ulster Television.

Luath Press Limited

committed to publishing well written books worth reading

LUATH PRESS takes its name from Robert Burns, whose little collie Luath (*Gael.*, swift or nimble) tripped up Jean Armour at a wedding and gave him the chance to speak to the woman who was to be his wife and the abiding love of his life. Burns called one of the 'Twa Dogs' Luath after Cuchullin's hunting dog in Ossian's *Fingal*.

Luath Press was established in 1981 in the heart of Burns country, and is now based a few steps up the road from Burns' first lodgings on Edinburgh's Royal Mile. Luath offers you distinctive writing with a hint of unexpected pleasures.

Most bookshops in the UK, the US, Canada, Australia, New Zealand and parts of Europe, either carry our books in stock or can order them for you. To order direct from us, please send a £sterling cheque, postal order, international money order or your credit card details (number, address of cardholder and expiry date) to us at the address below. Please add post and packing as follows: UK – £1.00 per delivery address; overseas surface mail – £2.50 per delivery address; overseas airmail – £3.50 for the first book to each delivery address, plus £1.00 for each additional book by airmail to the same address. If your order is a gift, we will happily enclose your card or message at no extra charge.

Luath Press Limited
543/2 Castlehill
The Royal Mile
Edinburgh EH1 2ND
Scotland
Telephone: +44 (0)131 225 4326 (24 hours)
email: sales@luath. co.uk
Website: www. luath.co.uk